# HELP! I need to
# MANAGE UP AND ACROSS

## 7 KEY INFLUENCE LESSONS

**TREVOR MANNING**

© Sept 2025

*Except as provided by the Copyright Act, no part of this publication may be reproduced, stored in a retrieval system or transmitted in any form or by any means without the prior written permission of Trevor Manning Consultancy Pty Ltd (trading as TMC Global)*

Designed and Typeset by JOSEP Book Designs
josework@gmail.com

Paperback ISBN 978-0-6454695-3-0
Ebook ISBN 978-0-6454695-4-7

# CONTENTS

Endorsements .................................................................. v
Acknowledgements ........................................................ ix

| | | |
|---|---|---|
| Chapter 1 | Busy Managing Down............................. | 1 |
| Chapter 2 | The Betrayal............................................ | 11 |
| Chapter 3 | Why bother Managing Up ...................... | 16 |
| Chapter 4 | The Interview.......................................... | 25 |
| Chapter 5 | Who's the Boss anyway? ......................... | 28 |
| Chapter 6 | The New Boss ......................................... | 35 |
| Chapter 7 | The Chameleon....................................... | 44 |
| Chapter 8 | Executive briefing disaster ..................... | 52 |
| Chapter 9 | The Battle ............................................... | 56 |
| Chapter 10 | Follow my Leader ................................... | 60 |
| Chapter 11 | Right Outcomes Despite my Boss............ | 69 |
| Chapter 12 | The Sales Meeting................................... | 78 |
| Chapter 13 | Map your Allies....................................... | 88 |
| Chapter 14 | How to Say No to my Boss...................... | 97 |
| Chapter 15 | Escalating the crisis................................ | 105 |
| Chapter 16 | Presenting Upwards................................ | 113 |

| Chapter 17 | Executive Briefing Success ..................... 122 |
| Chapter 18 | Gotcha ................................................. 125 |
| Chapter 19 | Transformation Complete ...................... 129 |

A Note from the Author ................................................. 135
Topic Index .................................................................... 137
Appendix 1 - Influencing Stakeholder Self-Survey ......... 139
Appendix 2 - Key Influencer Plan - Self-Reflection ........ 141
Recommended reading and References ......................... 145
About the author ........................................................... 147
Other Leadership books by Trevor Manning .................. 149

# ENDORSEMENTS

A must-read for technical and emerging leaders who want to understand how to navigate influence beyond their immediate team. This book blends a relatable allegory with practical leadership tools in a way that makes the lessons stick.

*Dr James Rink, Program Director, Leading US University, USA*

This book teaches leadership concepts through an engaging story using a multitude of learning styles: theory, relatable real-world examples, and guidance on how to apply the concepts.

*James Attree, Product manager, Software company, UK*

This book offers a fresh perspective on understanding challenges in networking within your firm to build relationships with co-workers outside of your reporting line. A great read for anyone looking to grow within an organization and understand people at different managerial levels.

*Rick Samson, Facilities Manager, US State University, USA*

Most leadership textbooks are dry and academic. This book is different—its engaging story brings influential concepts to life in a refreshingly authentic way.

*Sean Manning, Principal Solutions Engineer, International Software Company, UK*

An easy and quick read for mid-level managers looking to understand how their managers and *their* managers operate and how they, in turn, can support the up-and-comers by reading it and being refreshed as well. It would be a great resource to be spread and read widely through most organisations.

*Stuart McCullough, Head of Sales, Business Broker, Australia*

Why does no one recognize my team's work? Why don't leaders listen? If you've asked yourself these questions, this book is for you. It offers practical tools, real-world examples, and fresh perspectives to help you grow personally and professionally while tackling everyday workplace challenges.

*Dennis Adams, Training Manager, Electronic appliance manufacturer, USA*

This book is a challenging yet easy read, with guidance on how to understand others' perspectives and simplify and tailor the message to stakeholders.

*Neil Barnett, VP Product Development, International Agricultural manufacturer, Canada*

This book provides powerful alternative leadership insight, as well as offering practical strategies that extend beyond traditional leadership models. The distinction it draws between influence and authority resonated most with me, and has significantly shaped my approach to effective collaboration and leadership.

*Eddie Stephanou, Technical Manager,*
*International Wireless Manufacturer, Australia*

An enjoyable and easy read packed full of wisdom, practical lessons and moments that make you sit back and say "aha" as it all clicks into place.

*Mitch Hartas, Injury Management Specialist,*
*Queensland Police Service, Australia*

## ACKNOWLEDGEMENTS

There's a saying among authors: *a book is never finished, only published.* That rings especially true for this book.

The manuscript was completed in draft form over three years ago. But critical feedback from my beta readers encouraged me to do a complete rewrite.

My goal was to use true stories from my career to illustrate what I learned about having influence at work. Like many technical professionals, I once saw managers as people who existed to support *me*—not the other way around. Over a 30-plus year career, I've worked with many leaders—some inspiring, some deeply frustrating—and I wanted to share those experiences to help others avoid the mistakes I made.

To make the book more engaging, I enrolled in a fiction writing course to improve my storytelling skills.

I want to thank my wife **Berenice,** for her patience in editing this book with its endless changes and updates. Your challenging feedback and suggestions has resulted in a far better book than the original.

I'm especially grateful to **Kiera Young** for suggesting that I summarise the business lessons so they wouldn't get lost in the story. The seven-lesson learning journey approach was derived from that feedback. I have also included a content summary at the end of the book for people who want to easily find the main *managing up and across* concepts.

**Andrea Loppnow** helped me bring more authenticity to the characters. The revised book stays true to the people I've worked with—without compromising their privacy.

**Sean Manning** provided valuable insights on workplace diversity that helped shape the characters in the book reflecting generational, gender and lifestyle differences without judgment or stereotyping.

Thanks to **Stephen Dale** for his thoughtful feedback. Some of Stephen's own early-career experiences are reflected in the book, and his real-world perspective was very helpful.

I'm also grateful to **Alan Murphy**, who ensured the engineering elements would resonate with technical readers, and to **Dave Edwards**, whose sharp eye and clear thinking helped validate the core concepts.

Finally, thanks to **Vanessa and Mitch Hartas**, who helped confirm that the book's lessons are relevant even outside the technical world. Their support reminded me that influence is a universal skill, required in a broad range of disciplines, including medicine.

To my Advanced Readers and everyone who contributed:

thank you. This book is stronger, clearer, and hopefully more relatable because of you.

# BUSY MANAGING DOWN

Chris tossed about on the bed with a hundred things on his mind.

He looked across at his bedside clock.

3 am.

He tried counting sheep. When he reached 100, he gave up, picked up a notebook and tiptoed to his home office as his wife, Gwen, was sleeping soundly beside him.

He wrote a list of everything that was on his mind.

'No wonder I can't sleep,' he said, reviewing the list. 'Tomorrow is going to be another busy day.' He switched off the office light and crept back into bed.

◆ ◆ ◆

Chris Brown was the engineering manager for a medium-sized medical instrumentation firm in the mid-west USA. He was an engineer with an MBA and was technically very skilled.

Chris was a self-proclaimed family man. He and his wife, Gwen, had three school-age children, Evelyn (14), Liam (11) and Nathan (9), and he often spoke about them at work. When Chris was first promoted into management, the senior leadership team were concerned about the long hours he worked, and so provided him with a business coach, Sonya, to help him.

Sonya had helped Chris understand that he needed a mix of management skills and inspirational leadership to get the most out of his team. She helped him to set up his team with structure and operational governance, gradually fully empowering the team members and enabling innovation to flourish. Sonya had been coaching Chris regularly for over a year, and these days, they only met monthly.

Chris tossed over again in his bed. He had stopped worrying about everything on his list, but now his mind drifted to his conversation with Sonya. Earlier that day, he had called her for their regular catchup from his home office.

*... I'm getting on top of leading my team. Everyone is clear on their roles. They're aligned on their business outcomes. Our meetings run smoothly. Reporting is effective.*

*My problem isn't my team — it's my bosses. They don't understand what I do. They want more and more reports and keep asking me to cut costs, when what I actually need is more budget.*

*Half of these emails are from areas of the business that aren't even directly related to my job description. I don't want my team being distracted by other people's priorities. It reduces our efficiency.*

He cringed as he recalled complaining to Sonya that his boss, Rob, added no value to him.

*Did I overdo it? Rob pays her salary. Perhaps I should have toned it down. I just find it all so frustrating.*

With another sigh, he turned over again, noting the time. 3.50 am.

When the alarm went off at 6 am, Chris moaned loudly. Gwen stirred and asked, 'Another bad night's sleep?'

'Yep. My bosses are driving me up the wall, and I don't know what to do about it. Sorry, I can't discuss it now, though. I have to drive into the office. No working from home for me today.'

An hour later, he was dressed, had eaten, and was ready to face the long commute to work.

'Why do they think everyone has to be in the office on a Wednesday?' Chris complained to himself as he sat in heavy traffic. 'I'll probably spend half the day on video calls anyway.'

Chris found an all-day parking lot and walked to his office after downing his first cup of coffee at the local coffee shop.

He had settled in to read the copious emails in his inbox when there was a light knock on the door.

It was Madison, Chris's star performer in his team.

'Hi Madison. How can I help?'

Madison explained the situation. A legacy customer, Beta Engineering, had purchased software under an old contract that allowed unlimited use for all their engineers. It was a small company then, but subsequently had grown to hundreds of engineers. To upgrade to the current software version and benefit from all the new features, the new policy was for the customer to upgrade to a per-user billing model. This policy created short-term challenges for the technical team to convert to the new licensing software and was significantly more expensive for the customer, as the engineers were currently sharing the licenses off a network server.

Madison presented Chris with two options. In Option A, the customer would re-sign the agreement at an increased fee to account for the user growth, but would maintain their unlimited user model. In option B, they would force the customer to switch to the *per-user* licensing model, but with a significant one-off discount.

'So those are the two options. I like option A, as the conversion to implement option B will be a nightmare, but I wanted to check with you before committing that to the customer.'

'I agree, Option A is best, so go with that.'

'The thing is—'

Chris motioned with his index finger for her to pause as he looked at his phone. 'Sorry, one second.' he said, before answering the call. It was his colleague Dave Smedley, who ran the projects department. They both reported to the same boss.

'Hi Dave. I am finishing up a meeting. I can be there in the next 10 minutes. Yes, I understand it is urgent. As I said, I will be there in under 10 minutes. Okay, thanks. Bye.'

Chris turned back to Madison. 'Sorry about that. I agree, go with Option A.'

'The thing is, Quenton is pushing me to go with Option B. He says it has a better financial result in the long term.'

'Don't worry about Quenton. He is the finance director, and this is an engineering issue. Option A is definitely the preferred engineering solution. Thanks for your hard work on this. Good job.' Chris gathered some papers on his desk, poised to leave his office.

Madison hesitated. Chris sat down again and asked. 'What's troubling you?'

'Chris, there is another thing. This new request from *sales,* where we need to document all our customer conversations on the CRM system, is killing me. I have nearly a dozen accounts. If I have to document every conversation on the system, I will never get time to do any other work. I totally understand the need to record important decisions or conversations impacting other parts of the business, but every call?'

'I'm meeting Chuck later today,' Chris replied. 'I'll discuss it with him. I'm sure he will be okay with you only documenting the important interactions.'

Chris almost ran to Dave's office, realising he had taken longer than the ten minutes he promised.

'Chris, I need a solution this morning as we have a call at 2pm with Doctors-R-Us. You know how important they are to us. They escalated the issue to their senior management team, who are demanding answers.'

Chris and Dave discussed various options, and Chris promised to recommend a proposed solution by midday.

Chris was in the middle of brainstorming some solutions to Dave's dilemma with his technical team when his phone rang again. It was the caterer asking what pizzas he wanted for his team lunch. On the first Wednesday of each month, managers were encouraged to meet with their team over lunch as a team-building exercise.

'Please give us a mix of your best-selling options and include a few vegetarian ones. Oh, and please make one of the vegetarian options gluten-free, as one of my team members is gluten intolerant.'

After Chris finished the meeting, he asked Madison, his lead technical person, to share the solution with Dave. 'Please offer to join the customer call this afternoon to explain the technicalities of the solution, if required, and call me if things get out of hand.'

'Sure, Chris,' said Madison.

After running around attending various other meetings, Chris realised he was late for the team lunch, which was all set up in a spare meeting room.

'Thanks for leaving me the broccoli pizza,' Chris joked.

'Now you know how I feel,' Tracy said. 'I know they make gluten-free pizza bases with broccoli, but I like the meat feast, and yet you always order vegetarian for me. I'm gluten intolerant, not meat intolerant. It really annoys me. Is it *that* hard to order the right pizzas for your team?'

She paused, realising the rest of the team were staring at her.

Chris tried to make light of the situation, and the rest of the informal lunch meeting went well. The team was bonding well, and he made a mental note to find out how to keep Tracy happy in future. Despite the unacceptable way she had voiced her opinion, she had a point that he had not paid enough attention to people who ate different choices from his.

Chris was enjoying the social time with his team. When he checked his watch, he realised he needed to leave to meet Chuck, the sales director. Chris hated being late.

When Chris arrived at Chuck's office, his oversized, leather executive chair was vacant. He enquired about Chuck's whereabouts with one of the sales team.

They passed on the message that Chuck had left for the day. 'Try him on his mobile.'

◆ ◆ ◆

Chris returned to his office and called Chuck to explain Madison's issue with the CRM records.

'Chris, I want to see *all* the conversations the technical account managers have with my customers. If I relax the requirement to "important ones only", people will use that as an excuse not to use the CRM system. Sorry, they will have to work smarter, not harder, and commit time to updating the system.'

Chris tried to object, knowing how hard Madison was already working and trying to get Chuck to care more about his team's issues, but Chuck was unmoved.

Chris moved on to another topic.

'I'm checking to see if you have all our technical inputs for our customer catchup with Redlands Medical tomorrow.'

'We have a catch-up with Redlands Medical tomorrow?'

'Yes, Chuck, I emailed you the agenda and technical attachments for the meeting. It was all in the same email you used to accept the meeting, so I assumed you had read it.'

'Eh, no. Please send it to me again. What time is the meeting and where is it?'

'It's all on the agenda, Chuck.'

'Okay, fine. Look, I will top and tail the meeting with

some commercial discussions, and you can lead the technical discussion.'

'I have not included time on the agenda for commercial discussions. When we originally discussed this, you said it was purely a technical catchup, but you wanted to be there to foster the relationship. I have already agreed on the agenda with the customer.'

'It's all good Chris. Please let me have a few minutes in the beginning. The commercial issues won't take long.'

Chris left the call feeling dejected but put on a brave face and went to meet Dave to see how the other customer call had gone.

'It was a disaster.' Dave said. 'They are threatening to cancel the contract.'

'But why didn't you present the technical alternative we discussed?' Chris said.

'Well, I ran into our boss, Rob, before the meeting, and he told me that management had not approved your proposed option. Something about delivery risks and supplier delays.'

'But I've told management I have that under control. Why don't they stay out of technical matters? Did Rob say he has an alternative?'

'No, he said he would ask you to develop a better solution. He made some cryptic comment that integration with other industry software would become a critical criteria and that I would understand what he meant in a couple of days.

Something is brewing in the company, but he wouldn't discuss it further. I got the impression he regretted even making reference to it.'

Chris left Dave's office, shaking his head in despair.

*Off to the next crisis.*

# THE BETRAYAL

Chris sat at his computer. It was 3pm on Friday and he was working from home.

He hovered his mouse over the meeting invite. The last time the CEO called an all-staff meeting, they laid off 10% of the company. *Surely engineering won't be affected? We are the only department that actually creates true value for the company by designing new things, and we are critically short-staffed as it is. Demand has slumped recently. I bet it's manufacturing. Yes, it has to be manufacturing. I think we will be fine.*

Chris clicked on the link and joined the online meeting. The CEO was nervously making inside jokes with his executive team. Chris watched the other faces on his screen feign enjoyment. Eventually, the CEO got down to business.

'Welcome everyone. Thank you for joining the call today.

I have an important announcement to make, and I want everyone across the company to hear the news at the same time. For those of you calling after-hours, I appreciate it.

'I am sure you have all noticed that our order book has reduced recently.'

'Here we go,' Chris said out loud, quickly checking that he was on mute as the CEO prattled on about the market, stakeholders, profit, and the competitive landscape.

Chris checked his video image repeatedly to ensure his frustration and slight apprehension were not evident to others on the call.

'Well, the good news is, we are doing something about the situation. I am thrilled to announce today that we are merging with Medsoft.'

'Medsoft! But they are our arch enemy!' Chris exclaimed as the CEO continued.

'This merger will result in some changes, and has opened up exciting new opportunities for several people.

'Firstly, I am stepping down as CEO. I will oversee the transition for the next two years in an advisory role and then retire.'

Chris remained poker-faced. *So you sell us all down the river and then go and retire. Nice.*

... 'and I am thrilled to announce that Robert Jacobs will replace me as CEO, effective immediately.'

Chris sat up abruptly. *Rob Jacobs? That's MY boss. Who will take his place? I don't want a new boss.*

He peered into the computer screen, demanding answers as the CEO continued his speech.

'Gabriella Martinez, the Medsoft CTO, has been appointed as our new Chief Technology Officer and Thomas Armstrong, who was the Programme Manager with Medsoft, will replace Robert as the director of Engineering.'

Chris heard nothing after that. He had a thousand questions on his mind. *Medsoft? Our arch enemy? Why on earth would we merge with them of all companies? And who is this new guy, Thomas? How is a programme manager going to run a design team? He won't know anything about our projects, our process, our customers, our way of doing things. This is a disaster.*

The moment the CEO finished the call, Chris's mobile phone lit up. It was his boss. Chris's finger hovered over the flashing green symbol for too long before he answered.

'Rob. Hi. Congratulations. Wow. Big news today. I didn't see that coming,' said Chris in a neutral tone.

'Yeah Chris. It's all been top secret. This was a very complex merger, as you can well imagine, considering our competitive positions in the industry.'

'I'll say,' said Chris. 'The announcement mentioned the synergies in our product development lines. Rob, how did they know about our plans? What we are working on in engineering is very specialised.'

'Well, I'm sorry about all the secrecy, Chris, but we put Fred and Tracy on the *due diligence* team.'

Chris felt his anger rising but remained silent as Rob continued.

'Fred has been here forever, as you know. He has extensive industry experience and knows all our products in detail. We put Tracy on the team as we wanted a new recruit to give us a fresh, objective perspective. Both of them have been absolute stars.'

*Stars? Fred is my worst performer, and Tracy has one foot out the door.*

'Rob, you do know Tracy is considering another role in another company?'

'Yes, I do. Being on the merger team is one reason we managed to keep her in the company. She is young, smart, has great ideas and—'

Chris was no longer listening. *Rob clearly thinks Fred and Tracy are stars, and senior management obviously doesn't think very highly of me. I wasn't included in any of this, and now I have a new boss. What's his name? Thomas? A programme manager running engineering? This doesn't seem good to me.*

'Chris? I asked you how you feel about the changes.'

'All good Rob.' *It seems a bit late to ask my opinion now. You've made your decisions.*

'You don't *sound* good, Chris. I know this is a lot to take in. Sleep on it and see how you feel tomorrow. I will get Sonya to call you.'

'Sonya? Rob, I have been working with Sonya as my

business coach since last year and have consistently followed her advice. In my last review, we all agreed I made great progress in leading my team, and I wasn't even considered in any of these changes. What is the point of getting Sonya involved?'

'Chris, I agree you have done a brilliant job of both managing and leading your team. Your next skill to master is *Managing Up And Across*. This is about learning how to have meaningful influence with people who don't report to you. There are a lot of aspects of management outside your team that you may not even have thought about. You are valuable to us. I want you to work with Sonya to learn more about Managing Up.'

Chris knew exactly what to say. *Forget Sonya. I feel betrayed - betrayed by you, betrayed by my team and quite frankly, betrayed by our illustrious, now ex-CEO, who is sailing off to his retirement.*

Instead, Chris replied. 'Sure Rob, I'll talk to Sonya. Have a good evening and once again, congratulations.'

Chris ended his call with Rob and was about to walk through to the living room to talk to his wife, Gwen, when his phone rang again. 'A head hunter?... from where?... What is the name of the company?... Yes, I know I said last time I wasn't interested in new roles…

'Yes, I'm interested,' he said.

'Please put my name forward for the interview.'

# WHY BOTHER MANAGING UP

Chris sat waiting at his favourite coffee shop across the road from the office. Following their usual routine, he and Sonya planned to have coffee before their meeting. People were coming and going, collecting takeaways, eating donuts. They didn't seem to have a care in the world and even the smell of freshly ground coffee beans did not lift his mood.

Sonya arrived looking surprisingly cheerful.

*Another Judas in our midst. I wonder how much she knew of all these goings-on behind my back.*

'The usual?' Sonya called over her shoulder as she walked to the counter to order their coffees. Chris nodded without smiling.

They settled down to drink their coffees but the small talk was a bit forced, especially given the strong relationship they had built over the past year of coaching sessions.

Back in the office, Sonya got down to business. 'Chris, I got a call from Rob on Monday asking me to increase the frequency of our get-togethers for a few months. He wants us to meet weekly. He said you seemed very unhappy with the company changes.'

'Well, I feel like they don't value me. You've obviously heard about the merger. Apparently, Fred - a guy about to retire - is the big industry expert they used for the technical due diligence. And then they asked Tracy about our culture. You know my challenges with Tracy. She threatened to resign because she was unhappy with everything here. Apparently, the only reason she stayed is because they offered her a place on the merger team. And they didn't even talk to me about it. Is this how little they think of me? They didn't even consider me in any of their discussions.'

'Chris, Rob told me last night that they invited you to the *Digital Transformation* conference last year, but you were too busy. Rob says this is the industry's flagship conference, and at that conference all this merger activity kicked off. You sent Fred, and that's how he met all the relevant people and got drawn onto the Merger team.'

'Well, I had a real job to do. I sent Fred because he was dispensable. Who would run the team if I was busy at some industry hobnobbing event?'

Sonya paused and then replied, 'Do you recall when you got your first management role that you were sceptical about anything that was not task-related?'

Chris sat silently without responding.

'Previously, I gave you a book that explained all the aspects of managing and leading your team. I want you to read another book.'

Sonya handed Chris a book entitled *"A Leader's Guide to Managing Up."*

'Read the first chapter, and when you're ready, let's talk again.'

*Managing up? I am not one of those people who wants to kiss up to management and lick their boots.*

'Sonya, if management can't appreciate my value, I certainly don't plan to go boasting about what I do to impress them. My work should speak for itself.'

'It's not about kissing up or boasting. It is about being proactive in the way you influence others. Nobody's work speaks for itself, especially if the people evaluating it are not experts in your tasks. If you stop something bad from happening, how would a manager know the value you or your team have added unless you inform them? Informing others about what you do and why it matters is not boasting. It is an essential part of building your credibility and business value so that others are happy to invest time and money in you.'

Sonya opened the book. 'Here is the learning journey we'll embark on.' She walked him through the chapters that were illustrated in a diagram.

# HELP! I NEED TO MANAGE UP AND ACROSS

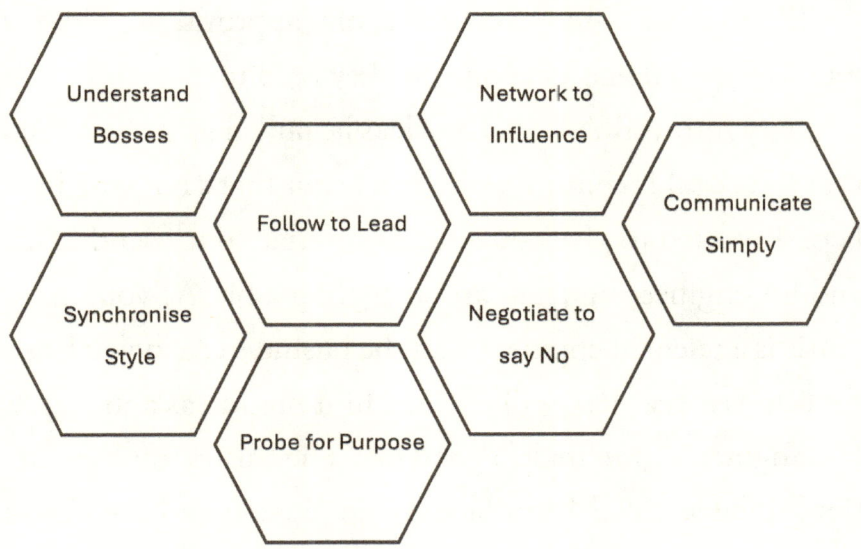

Chris looked dubious but could see Sonya was not backing down. He flicked through the book and said, 'Sure, Sonya, I'll read the book.'

It had been a few days since the announcement, and Rob was sitting in his office trying to come to terms with his new CEO role. It had all happened so fast and there were so many moving parts. The weeks leading up to the announcement had been chaotic, and the secrecy had made it very stressful. There was a loud knock on his door. Rob looked up. Filling the doorway was the imposing presence of Chuck, the Sales Director.

'Hi Rob. Got a minute?'

'Sure, what's up? You want my approval to close a multi-million dollar deal on your key account?'

'Very funny, Rob,' said Chuck as he pulled up a chair. 'Yes, that sales deal is coming along nicely but that's not why I am here. I wanted to give you a heads-up that your friend Chris and his engineering team are struggling a bit. As you know, Chris is a talented engineer, and the business coach you hired for him last year has really helped him mature as a manager. His engineers *want* to do a good job. They also want their designs to be *perfect*. My problem is, our customers have urgent requirements, and I'm trying to get the team to understand they have to balance getting the design perfect with getting it released. They never commit to deadlines. Their attitude is, *It will be done when it's done.* I need Chris to be more assertive with his team to meet the customer's requirements and help us achieve our numbers. I am not asking you to do anything about it. I have it under control. I wanted you to be aware that Chris is struggling, that's all.'

Rob reflected on what Chuck had said.

'Maybe you can help me with something, Chuck. The last time you told me that Chris was struggling with his current role, you also said that he had told you any promotion offers would put him in an awkward spot. You said Chris would feel compelled to apply, but had confided in you that he wasn't ready.'

Rob looked at Chuck for a reaction.

# HELP! I NEED TO MANAGE UP AND ACROSS

'That's right. Chris said he can hardly cope with the current pressure, let alone the pressure of a new role.'

Rob replied. 'That's odd. Chris is upset about not being considered for the Director of Engineering role in this merger.'

'Chris is a nice guy, but he is not director material.'

'I think he has some lessons to learn about his broader role in the company, but he is someone I have plans for in the business.'

'Well, you're the CEO now, Rob. It's your decision. I am just trying to help. And hey, please say nothing to him. He spoke to me in confidence.'

'I hear you,' replied Rob.

'By the way, the guests for my holiday property cancelled at the last minute. A sick child. So, the cabin is empty this weekend. I know being CEO is a tough gig, so if you'd like to take the family up there for the weekend, please shout.'

'Thanks Chuck. That sounds lovely, but I don't want to impose. Perhaps another time.'

'*Impose?* You'd be doing *me* a favour. I hate the place standing empty. I'll drop the keys off in your office in case you change your mind.' Chuck waved his hand in the air and converted it into a thumbs-up. 'Great to have you as our new CEO, Rob. Goodnight.'

Back in the Brown household, Chris sat down to read Chapter 1 of the book Sonya had given him.

## INTRODUCTION TO MANAGING UP

The role of the modern manager is not easy. Constant technological advancements render their expertise potentially obsolete while financial constraints tighten and performance expectations escalate. Concurrently, employees seek purpose, striving for a harmonious balance between professional and personal lives. Furthermore, navigating ambiguous and flexible hierarchies and virtual teams adds another layer of complexity, where managers are not sure of who their actual bosses are and what real control they have over their own staff.

Managers often possess firsthand knowledge of the team's tasks, having previously fulfilled those roles. However, understanding the role of their own superiors remains elusive, as they have not occupied such positions. Despite this, managers often negatively evaluate their superiors, particularly when they perceive minimal personal benefit from the relationship.

Managing one's boss may seem unconventional, yet neglecting this aspect can undermine effectiveness in addressing the non-tangible demands of senior management. These demands, often obscured by various factors, require deciphering true intentions beyond explicit requests. There is a skill in determining what someone actually wants rather than what they have asked for.

Modern organisations operate with a dual structure – a structured hierarchical one with rigid systems and processes designed to eliminate individual autonomy and reduce risk, and a second invisible structure more akin to a start-up company, which fosters growth, agility, innovation and autonomy. To thrive in this dynamic environment, employees must adopt an executive mindset, a concept pioneered by Peter Drucker with the advent of *Knowledge Workers*. Arguably, we now live in a post-knowledge era workplace with Artificial Intelligence (AI) challenging the value of human expertise. The future success of managers lies in the fact that computers may take over the routine, process-based functions, and the success of managers will lie in their capacity to get results in an environment of contradiction and complexity.

Learning to positively influence others and adapt company structures and processes up and across the business is a critical human skill to master. Managers who do not build robust and influential networks of people up and across the business, both inside and outside their organisations, will find it hard to be effective in this new dynamic, complex world of work.

◆ ◆ ◆

Chris shut the book.

*All this theory. I assume Sonya thinks I am the problem because she insists I read this book. Why do I need to get involved in managing relationships up and across the business? Why can't my bosses improve the way they deal with me?*

# THE INTERVIEW

Chris checked the address supplied to him by the recruitment agency. Satisfied that he had the right place, he stepped into the elevator, usually dedicated to the executives only, and pushed the button to the top floor. The door closed heavily behind him. After a long, slow ride, the door opened again. Glass windows trimmed with gold plating extended from floor to ceiling, revealing expansive views across the city. A massive mahogany table polished to perfection sat in the middle of the room. Black leather chairs guarded access to the table. *Only important people are allowed.* The room was cold, with a metallic smell blended with wood polish. Dimly lit photographs of previous chairpersons, gloomily framed in a dark wood, filled the walls. The picture of one female executive stood out, with her kind, broad smile and bright red dress, breaking the mould of the dark grey pinstripe suits and poker faces. The atmosphere was awe-inspiring, yet Chris felt alone, an intruder, an unwanted peasant. He inspected a trophy on display. It was cold, empty, and threatening. *Go away*, it declared. *You don't belong here.*

A door opened and three stern faced people walked in with clipboards. The woman who appeared to lead the group spoke first. 'Welcome. Um–' She looked down at her notes. 'Chris. Please sit.'

They placed Chris at one end of the table, with the three of them sitting opposite like a police interrogation.

'First, I must caution you that anything said here is strictly confidential. We may share company information as part of the interview process, and we expect you will reveal nothing we share outside these four walls. Understood?'

Chris nodded without speaking.

'So, could you tell me what your greatest weakness is?' All three assessors raised their pens, ready to record Chris's every word.

After 30 minutes, Chris could see they were only halfway through the questions on the template and asked for a glass of water.

The water arrived, and the cold glass perfectly summed up the atmosphere in the room. Finally, the meeting ended, and Chris returned to the mirrored elevator.

Chris was alone and looked at his reflection in the mirror. *What are you doing? Maybe you need to make things work out where you are.*

# WHO'S THE BOSS ANYWAY?

## LESSON 1 - YOUR LINE MANAGER IS ONLY ONE BOSS - NAVIGATE THE MATRIX!

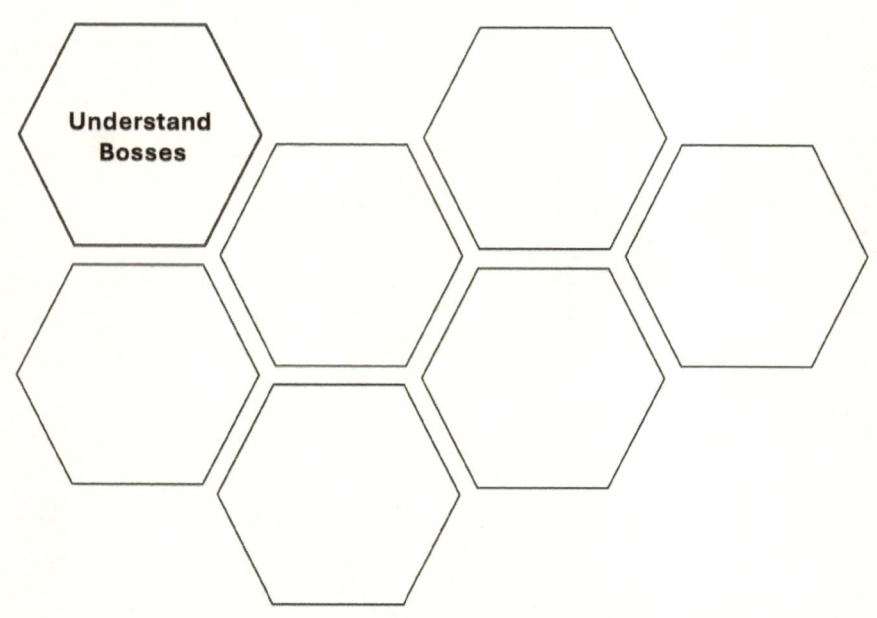

Chris sat opposite Sonya, stirring his coffee despite not adding sugar. A week had passed since they last spoke.

'Sonya, my question is, why do we even *need* bosses? I can handle my job perfectly well, as can my team, without help or input from my bosses. I don't see them contributing anything to me or my team. In fact, I don't know what they do. I reckon that you could fire the lot of them and all that would happen is our profits would increase.

'No matter what we do, it's never good enough. They sit around in meetings all day and then come out and upset all the progress we are making, with contradictory priorities and unreasonable demands.'

'Chris, to ask why we need bosses is actually the wrong question and highlights the mind-set problem in many organisations. The real question is why we need the team. Remember how Larry Page from Google fired all his non-technical managers in 2001? It wasn't long before he hired them all back. Bosses are necessary to set direction, assess and communicate progress, and help with alignment. They create order and structure to allow for sustainable growth.

'Imagine a start-up business with no defined roles. Who is responsible for strategy? Who is responsible for getting the product out on time? Who is responsible for customer relations? Who is responsible for banking issues? Who is responsible for the leaking toilet? The list goes on and on. The founder is responsible for everything.

'The functional management structure is there to *own* functional responsibilities. The finance manager owns anything to do with finances. The operational manager runs and owns the day-to-day business affairs. The sales and marketing manager is responsible for building a market for the product and owns sales orders. With a functional organisational structure, unique responsibilities are clearly defined: when the toilet leaks, the founder no longer needs to get involved; when sales are down, the sales manager presents the recovery plan; and when the bank asks about refinancing a loan, it is clear who will talk to them.'

'Functional ownership makes sense, but my frustration is the lack of teamwork across the company. Everyone seems to only look after their own Key Performance Indices so they qualify for their bonus. Departments often achieve their own KPIs at my department's expense.'

'Well, in that case, it might be a good ideas to review the KPIs so there is alignment. Imagine a football team with no unique roles. Everyone would be well meaning in trying to score goals, but it would be chaos. If KPIs are not aligned, part of managing across is to ensure they become aligned.

'Chris, you mentioned earlier that you don't know what your bosses do. Of course you don't - you have never done their role. Part of *managing up* is to find out. There is a chapter in your Managing Up book about what bosses actually do and what keeps them awake at night.'

Later that evening, Chris settled down to read.

## WHAT BOSSES DO

Bosses do whatever is necessary to meet the demands of *their* bosses. Typically, senior managers undertake the task of strategic planning and forecasting potential outcomes. Their primary concern is envisioning the intended results, often overlooking intricate details. Within their sphere of influence, they delegate the minutiae to the teams under their supervision.

## WHAT BOSSES WORRY ABOUT

Bosses worry more about extrapolating the general case than any individual result or event. A departmental manager needs to build processes to achieve predetermined outputs like a franchise agreement. For instance, if a customer received a charred hamburger from a McDonald's outlet, the franchise owner's foremost concern would be the systemic flaw that permitted the burger to burn rather than the isolated incident itself. While a singular occurrence may constitute a regrettable mishap, a recurring flaw in the process could jeopardise the viability of the entire enterprise. Consequently, while acknowledging the immediate grievances of a dissatisfied customer, the paramount concern for the business lies in rectifying the systemic deficiencies of the general case.

Although bosses may delegate specific responsibilities to subordinates, they ultimately bear sole accountability when reporting to higher authorities. Occupying a distinct hierarchical position within the organisation, bosses assume a unique mantle of responsibility for specific outcomes. Failure to delineate and uphold these responsibilities often results in a diffusion of accountability, wherein multiple individuals are perceived as responsible, yet no one assumes definitive ownership. This organisational ambiguity fosters redundancy and oversight gaps, wherein tasks are either duplicated or left unattended due to assumptions regarding the responsibilities of others.

## NAVIGATE THE MATRIX

In the past, bosses operated as supervisors with a superior knowledge of the tasks and so could ensure proper execution. A clear hierarchy was in place, with workers following directives without question. Bosses held hiring and firing authority, with staff solely accountable to them. Accessing team members without the boss's consent was not condoned, and bosses shielded their team from conflicting demands.

Today's organisational landscape differs markedly. Structures are flatter, with cross-functional teams and dotted-line reporting. Numerous individuals outside the direct reporting line wield influence over workloads and priorities. Digital connectivity enables anyone within the organisation to assign tasks to any management tier, disregarding traditional boundaries and job scopes. Remote work exacerbates managerial disempowerment. Managers unable to observe their team working hinders oversight and immediate intervention. Additionally, digitisation erodes information barriers, granting junior staff access to the same data as senior management.

So, who is the real boss? The organisational chart usually defines a short-term home for team members based on the functional definition of their current role. The person in charge of that function is recognised as the team's line manager. They have the authority to control the things defined as part of that role, but they do not necessarily have the most power in the organisation. The real boss is the person who keeps the team member hired. The real boss defines the priorities of the line manager. The real boss looks after the team member's long-term career.

Chris closed the book and switched off his reading light. *This is starting to make sense. Is that why Rob wasn't as concerned as I thought he should be about the Kramer account? I presented it to him as a burnt hamburger. He is responsible for dozens of accounts, so he seemingly didn't care. Next time, I need to lead with the common thread that affects all the accounts. That will get his attention.*

The next morning, he woke up early and headed off to work. He had a ton of work and a day filled with meetings. Chris noticed that Chuck's car was already there. *That's unusual. Chuck is usually the last to arrive.*

Chris thought he would be friendly and popped his head into Chuck's office to say hello. *I mustn't get stuck there. I came in early to get some work done.*

Chuck's office door was slightly ajar. Chris pushed it open and offered a cheery greeting. 'Morning.'

Chuck looked up at Chris. 'Have you never heard of knocking?' Chuck scrambled some papers together and pivoted his screen out of view. 'What can I do for you?'

'Nothing, I was just saying hello.'

'Oh, okay. Sorry, I am working on an important deal with a deadline. Do me a favour and close the door behind you when you leave.'

Chris walked to his office, deep in thought. *What are all these confidential deals that Chuck keeps referring to?*

# THE NEW BOSS

Sonya sat down in an uncomfortable visitor's chair. Chris's new boss, Thomas, had directed her into her seat, then took a call on his mobile phone and left the office, whispering to her, 'Sorry, I won't be long.' That was ten minutes ago.

She looked around Thomas's office. It was neat as a pin. He had ice hockey trophies and photos from his college days proudly displayed around his office. *He is not as fit as he used to be.* On the side wall was a set of framed certificates boasting his various qualifications. Next to them was a picture of him shaking hands with Mike Tyson. *That's a strange hero?*

Thomas marched back into his office, sat on his large leather chair, and leaned backwards. 'Sorry about that. Important business.' Thomas leaned forward, placing his forearms firmly on his desk as he spoke. 'Look, Sandra, let me be straight—'

'My name is Sonya.'

Thomas looked down and studied Sonya's business card carefully before responding. 'Sonya. Look, I am a straightforward guy and I like to be open and honest with people. I don't like business coaches. Don't take it personally. I think *those that can, do, and those that can't*—well, you get my point.

'I think they selected me to whip this place into shape.'

Thomas pointed to a photograph on the wall that the previous occupant had left behind. It was of the pre-merger management team.

'We have been competing with *these people* in the marketplace for years and they are too slow, too soft and too conventional. If I had my way, I would replace Chris to attract fresh talent, but fortunately for him, his old boss rates him highly and is now the CEO and so I have no choice but to keep Chris.

'In my old company, MedSoft, we only tolerated A-players. I hope our merger doesn't mean we are going to drop our standards. Anyway, Rob told me he wants you to continue working with Chris as his business coach. That's fine with me. One less thing for me to worry about, but let me be clear. I haven't got time to hold his hand, and to be perfectly frank, I don't have time to have lots of touchy-feely meetings with you about him either. Are we clear?'

Sonya stood up and reached out her hand for a handshake. 'Perfectly clear.' She smiled graciously and left his office.

Sonya had planned to meet Chris at the coffee shop after her meeting with Thomas, so she sent a text to see if he could meet earlier than originally planned.

Chris replied. 'Sure, for you, I always have time. I'll see you downstairs in five.'

'I'll order this time,' said Chris as they arrived at the coffee shop. He ordered 2 coffees and then sat down.

'So you've met Prince Charming?' Chris asked. 'I am not sure we are going to get on. He clearly doesn't rate me and wasn't interested in any of my ideas to improve the department. I offered to run him through all our product plans in detail so he could get up to speed but he turned me down. Said he wanted to stay high level, and that it was my job to worry about the details. Said he didn't have time for lots of talking. Said he wanted to see actions. Do you know it was two days before he even came over to introduce himself?'

'Chris. I get it. He has his style, you have yours. Did you think of going over to introduce yourself to him?'

Chris looked into his cup.

Sonya let it go and asked, 'How is the book on Managing Up coming along?'

'Yes, good, but it implies it's all down to me. It's my responsibility to manage up. It's my responsibility to deal with all these new people across the business. I bet the author of the Managing Up book thinks it's my responsibility to deal with the incompetence of my new boss.'

Sonya looked disapproving. 'Chris, what outcome do you want at work?'

'I want things to go back to the way they were, but I guess that ship has sailed. Look, this is a great company and I think Rob will make a great CEO. MedSoft, the company we merged with, has impressive new technology, which we now have access to in-house. I think if we combine our experience and expertise with the fresh new thinking and technology of our merged company, our new business will be hard to beat in the marketplace. I'd like to be part of that. I have no idea how I can get around the fact that my previous boss betrayed me, my team betrayed me, and now I have a new boss who hates me. Can you imagine how hard it will be for me to report to him?'

'Chris, have you ever thought what it would be like to manage *you*?'

Chris sat silently for a solid fifteen seconds.

'You know, no, I have never considered that. I have been so busy thinking about how all this affects me that it hasn't crossed my mind to consider what it's like for everyone else.'

Chris was silent again.

Sonya continued. 'By the way, there is something that has been puzzling me. I was told something in confidence, so if I discuss this with you, I need your reassurance you will keep it that way.'

'Sure.' Chris looked intrigued.

'Rob told me he had a confusing conversation with Chuck

this week. Apparently, you told Chuck you didn't want the director role. You felt you weren't ready for additional responsibility. I was surprised to hear that, and Rob is confused about why you reacted so strongly about not getting the role when you didn't actually want it.'

'What? That is a complete lie. I have never had any such conversations with Chuck. I need to talk to Rob and explain my side.'

'Chris. I would not advise that. Rob told me in confidence. This will be your word against Chuck's. I don't know what Chuck is up to, but you might need to be careful when dealing with him. Keep a written record to protect yourself from harm. And Chris, do not go on the attack. If I am right about Chuck, he is the type of person who will always be one step ahead and manipulate the situation to his favour. You will lose if you try to confront this head-on.'

'I can't believe he would do that to me. I certainly will be extra careful in my dealings with him.'

'Chris, put the Chuck issue to one side. Let's work on your issues with Thomas.'

'I think I need to talk to Rob and point out what a bad manager Thomas is. I think it was a terrible decision to bring an arrogant and incompetent person like him in as engineering director. He doesn't have any clue about our products and that's why he wants to stay high level.' Chris punctuated the words *high level* with imaginary apostrophes in the air.

'I am worried about the effect on the company's success. We have always prided ourselves on our outstanding product quality. I really think Rob will see sense if I talk to him.'

'Chris, who do you think appointed Thomas?'

Chris groaned and put his head in his hands. 'Well, what do I do then, Sonya? I can't stand by and let this new guy ruin everything we have built so far. I have had zero success even starting to have any influence over Thomas. You are saying my great historical relationship with Rob won't help when it comes to Thomas, so what do I do?'

'If you want to influence, start by building a relationship with the person you want to influence and start with their issues, not yours. You say you don't want Thomas to ruin everything you have built. Do you really think Thomas wants to destroy it?'

'I suppose not. But, by not understanding our product direction, with his arrogant attitude and his power from the authority he has been given, Thomas could ruin it before he even knows he is doing it.'

'Your challenge is to help him not to. Start by building an influential relationship with him. Imagine what it's like being him and help him with the issues he is likely facing.'

Sonya pulled out her copy of the Managing Up textbook and showed him where she wanted him to read. 'There's good advice in there that should help you with your new boss.'

After the meeting with Sonya, Chris headed back

to the office and made an appointment to catch up with Rob. He now realised why Rob had vetoed his solution for Doctors-R-Us that he had given Dave. Armed with the new information about the merger, he felt sheepish about how he had judged Rob.

Chris had a good meeting with Rob and was able to come up with an acceptable alternative to share with Dave.

That night, Chris picked up the book and started reading the section Sonya had referred him to on New Bosses.

## NEW BOSSES

Getting a new boss is challenging for both parties. The appointment of a new boss can stem from various catalysts, ranging from the predecessor's retirement, promotion, or dismissal. Consequently, the arrival of a new leader heralds a shift in expectations and dynamics within the team. For the incumbent team members, the advent of a new boss elicits a dual response, embodying both trepidation and promise. Relationships forged with the former leader may evoke sentiments of camaraderie and protection in some, while provoking relief and anticipation of change in others. However, for a select few, the transition may stir feelings of disenchantment or overlooked potential, amplifying sentiments of exclusion or discontent.

New bosses will inevitably have a mixture of emotions as they handle their new assignment. Drawing from past triumphs, they aspire to transplant successful strategies from prior roles into their newfound domain. The problem is the new domain may have unforeseen differences.

Lack of knowledge about their new environment can be overwhelming, notwithstanding the reasons for promotion. They may have been the go-to-person who had all the answers to everything. In their new role, they probably do not even know the names of their team members, let alone understand their strengths, weaknesses, and abilities. In the beginning, it may be hard to understand the complexities of actually *implementing* the company strategy that attracted them to take on the role. The new boss may have a sense that they are literally clueless about their current environment.

Conversely, being managed by a new boss is not easy either. Often tasked with orienting the incoming leader, team members assume a quasi-mentorial role, leveraging their localised expertise to bridge the knowledge gap. This transient phase may blur hierarchical lines, imbuing team members with a semblance of authority over the new boss, potentially engendering friction, particularly if the promotion bypassed deserving candidates within the team.

To influence a new boss, aim to help them. Make the first move and help to integrate them. Senior management will expect a new manager to change things. By aligning with the change agenda, individuals position themselves as catalysts for progress, safeguarding against alienation and marginalisation from decision-making spheres.

People who genuinely can mitigate the risks of change often find themselves outside of the process with no influence and are unable to de-risk the changes. They become excluded because it is perceived they are against any change. They are not seen as an enabler of change. It is preferable to embrace change and become a champion of transformative initiatives. In cultivating a constructive outlook, they exert influence from within and can steer the changes in the right direction. Better to be inside the change process than on the outside with no influence.

Chris closed the book sheepishly. *Once again, I haven't been considering this from Thomas's perspective. Looks like I do have something to learn about Managing Up.*

# THE CHAMELEON

## LESSON 2 - SYNCHRONISE YOUR STYLE TO THAT OF YOUR BOSS

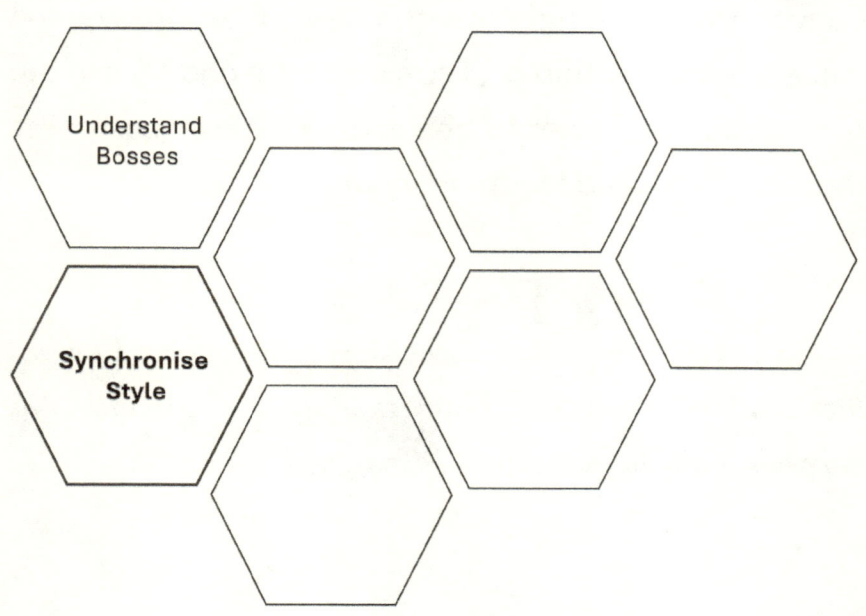

Over the next few weeks, Chris tried really hard to look at things from the perspective of other people. It was quite eye opening, and he was pleased with how it improved his relationships. What confused him was that it hadn't improved his influence. He talked to Sonya about it in their next meeting, over lunch.

'How is your Managing Up journey going, Chris?'

'I can definitely see that by focusing on the other person's perspective, my relationships have improved. What hasn't improved is my influence. I value having easy-going relationships, but my job requires me to influence business outcomes. It is not a popularity contest.'

'Can you give me an example?'

'Well, I did a needs analysis of the various stakeholders to look at things from their perspective.

'Rob, as CEO, wants the company to meet the sales number for the quarter, but he also wants long-term customer loyalty.

'Olivia, as the HR director, wants high overall employee engagement, and I'm sure she also wants to avoid employment tribunals caused by Chuck.

'Quenton, as Chief Financial Officer, wants to save costs and meet the profitability targets.

'Gabriella, as Chief Technology Officer, wants to release appealing new products to the market.

'And Thomas wants to make a good impression as the new Head of Engineering and be effective in his job.

'Next, I did a detailed analysis of the situation. I even created graphs that illustrated the points I was making. I then sent them all separate emails specifically dealing with the issues from their perspective. I put a lot of effort into it. I sent these off a few days ago, but I have had no response, well apart from one, which I received just before this meeting.

'Let me show you.' Chris took his phone out and scrolled to the email reply from Quenton, the CFO. Quenton had used a different font colour to respond to each point that Chris had made. Most of Quenton's comments were pedantic, even occasionally correcting Chris's spelling.

'Chris, how would you describe your preferred working style? How do you like to communicate and collaborate?'

'That's easy, Sonya. I like structure and clarity. I like to work on my own and think things through deeply, then write it all down so it makes sense. I don't like sharing my ideas until I know they are logical and can be implemented. I analyse any risks so there are no surprises during implementation. I enjoy sharing my ideas through email because I am not pressured to think on the spot. Using email, I can carefully consider what I want to say.'

'And what is Quenton's style?'

'Like me, he prefers to communicate by email and likes to critique the fine details. I have seen him in meetings correct someone who has put the wrong date on a PowerPoint slide, even if that date is irrelevant to the conversation. My beef with him is that he only cares about the numbers. I'm trying to get him to care about the technical details.'

'Quenton is aware that you are the technical expert, and you've said he likes to critique the content. By sending him a technically focused document, Quenton adds value in the only way he knows - to correct any errors in numbers, spelling or layout. If you want his attention, start by sharing the financial impact of your technical point. The only way he'll take the technical details seriously is if you show him how they affect his numbers.'

'Good point.'

'And what would you say Rob's style is?' Sonya asked.

'Rob? He is a people person. His first thoughts are about my personal wellbeing. Instead of critically analysing what I am telling him, he always explains how it will affect morale. Sometimes it really frustrates me.'

'So how did your email address Rob's concerns about people?'

'It didn't. But I wanted Rob to have a better understanding of the technical aspects.'

'In what way do those technical details impact Rob?'

'Well, if we don't get a successful product launch, we will end up laying people off. Surely Rob cares about that?'

'Of course he does. Did your detailed email explain to

Rob that this is at least partly about future opportunities for the people?'

'Well, no, but I would have hoped he could join the dots.'

'Chris, Rob is a busy guy. There is a reason he didn't respond to your email. One of them is that he is not an "email guy". Secondly, your email didn't appear to address the issues that are top of his mind. If you want to increase your influence, make it easy for the other person to see how your issue affects them.'

'And Gabriella?' Sonya asked?

'She is an entrepreneur. Very unstructured but very creative. She talks a lot and can change her mind from day to day without batting an eyelid. Gabriella seems to live in her own little product world. She seldom concerns herself with implementation or costing details. She wants to design beautiful solutions that sometimes don't make financial sense and are hard to build. I suppose I was hoping my email would get her to focus on the practical implementation issues. But now that I think about it, an email about improving implementation efficiency will not get her attention. I would probably be better off discussing this with Thomas and then arranging a joint meeting to show how better implementation achieves the goal of more exciting future products.'

'Good insight. And what about Olivia, your HR person?' Sonya asked.

'She is very strategic. She seems to live in a parallel universe focused purely on the future. She doesn't really understand the business operationally, nor does she want to. She seldom gets riled about any day-to-day issues, which is great at one level, because she is always calm and collected, but she lacks urgency on issues that need immediate attention. I bet she hasn't even read my email, and even when she does, I doubt she will look at my analysis.'

'Chris, why did you include analysis if you know she won't read it?'

'Because I thought she *should* read it. It is important for her to understand the operational aspects of the business. Do you know she doesn't even know how to log onto our CRM system, where all our sales and engineering reports are? Every time she asks me about some operational issue, I tell her it's all in the system, and I have shown her countless times how to access the reports, but she finds it easier to ask me to do it every time. She *needs* to learn.'

Sonya said nothing.

Chris grinned. 'Listen to me. I'm doing it again, aren't I? Looking at things from my perspective, not theirs.'

Sonya smiled back. 'As a matter of interest, how would you characterise the style of your new boss, Thomas?'

'Thomas? He is a taskmaster. He thinks the best way to get things done is to bang his fist on the table and demand them. Since he joined, we have had countless run-ins because he doesn't read my emails and then demands verbal answers from me in meetings, embarrassing me in front of the team when I didn't know the answers off the top of my head.'

'Chris, can you see the pattern here? You want everyone to engage with you in your preferred working style. If you're going to influence them, not only do you need to address their needs before your own, you also need to engage with them in *their* preferred style.

'Olivia is not going to learn your operational system. She wants to have a conversation, which will likely be about the long term.

'Rob will not read through your detailed analysis. He will want to discuss how what you are proposing affects the company's people from a top-down perspective.

'Quenton will love receiving emails from you instead of a conversation, but you will need to be much clearer about how you want him to respond, or he will see his role as auditing the accuracy of your work.

'You need to be a chameleon, Chris, adapting to each person's preferred style.'

'Chameleon?' Chris asked. 'You're kidding, right? What happened to authenticity? I thought we were supposed to bring our authentic selves to work?'

'Absolutely. Authenticity is important. This is another one of those management contradictions. At one level, it is essential to be authentic because it builds trust, which is a key ingredient of strong relationships. On the other hand, your authentic self may want to arrive at work in shorts and a T-shirt and play computer games all day. Your authentic self probably wants to tell your customers to take a hike when they give you a hard time. Your authentic self often wants everyone to accept you just the way you are and makes no effort to adapt to their preferences. In the real world, we need an adaptive self. When with field staff, talk their language and engage in their world to get the best out of them. When you are with senior managers, conform to their norms of behaviour to have the most influence.'

With the uneaten olives strategically hiding under his knife and fork, Chris pushed the remains of his salad lunch aside and looked straight at Sonya.

'A chameleon? I have to think about this one, Sonya.

'Thanks for lunch. Got to run. I have a presentation to make at the monthly senior management meeting. I wish I had had this discussion before preparing my presentation, but we'll see how it goes.'

# EXECUTIVE BRIEFING DISASTER

Chris walked into the meeting room.

Rob, the newly appointed CEO, sat at the head of the table, with Olivia, the HR director, sitting to his right. Dotted around the table were the other senior departmental managers, including Quenton, the CFO.

Chris had hoped to see Gabriella, the new CTO, in action at the management meeting, but she was attending a customer meeting with Thomas, his new boss.

A half eaten plate of pastries sat tantalisingly in the middle of the table, with empty coffee cups strewn about untidily. The room smelled a little musty. *It must be a long meeting.*

Rob looked up at Chris. 'Welcome. Find yourself a seat and help yourself to a pastry. They are delicious.'

Rob waited for Chris to settle, then asked. 'So what's next on the agenda?'

Quenton spoke up. 'Chris is going to update us on the progress of the latest product development plans, and we need to decide if we should increase funding for these projects.'

'Okay, Chris, the floor is yours,' said Rob.

Chris started sharing the PowerPoint presentation he had hurriedly put together the night before.

'Sorry, Thomas only told me a few days ago that he wanted me to do this update and I have been flat out. I am not as prepared as I'd like to be.'

'Yes, we've had a lot going on here too,' said Quenton without looking up.

'I'd like to start by sharing our problems with the current software,' said Chris. He then presented a detailed slide with multiple columns of data.

'As you can see, well, it might be too small to read, but the table shows latency issues and response times are below the industry benchmarks. The high standard deviation from the norm clearly demonstrates the source of our problem.' Chris looked up.

Rob nodded enthusiastically, willing him on, yet his eyes showed he had no idea what he was talking about. Olivia, the HR director, looked serious and was texting on her phone under the desk. Quenton was copying down data furiously and updating a spreadsheet. *Is he correcting my PowerPoint slides?* The other managers looked on, expressionless.

Chris moved on hastily through his 30-slide pack. He was asked to limit his presentation to 30 minutes at most and so he felt the need to rush through it . After 30 minutes, Quenton stopped him mid-sentence while he was explaining the complexities of a new code release and said, 'Chris, we only have 15 minutes left to make a decision. Can we stop you there, please?

'Before Chris leaves, any questions?'

Olivia spoke up. 'I noticed you were advancing the slides with your mobile phone. How do you do that?'

'It's an app I have. I can show you afterwards if you like.'

'Okay, thanks. Sorry. I hadn't seen it before. No, I have no questions about the presentation.'

Rob spoke up. 'In that case, thanks, Chris. I'll update Thomas with our decision.' Rob smiled kindly as Chris left.

Meanwhile, Tracy had been catching up with her colleague Madison, one of Chris's most experienced team members.

'Madison, I want some advice from you. It's about Chris. I know he's in the management meeting, so thought it would be a good time to talk.'

'Sure, what's up?'

'Well, I can't seem to get Chris to act with any urgency. I

told him about a problem I am having with my customer, who is complaining about the speed of the software, and Chris keeps fobbing me off.'

'What do you mean by *fobbing me off*?'

'Well, he tells me I have to improve communication with the customer, so they understand our roadmap. It's not about the roadmap. The customer wants improvements in speed now, and Chris won't commit to anything. I can see the customer's point of view. Our software is too slow!'

'I think we all agree we have performance issues with our software. There are plans in place to improve future software releases significantly, but it all takes time. It could be that Chris doesn't want to over-promise.'

'Chris is too conservative,' Tracy said.

'Yeah and Chris also has a lot of experience. Have you tried to understand Chris's perspective on this?'

'No, and I don't plan to. They chose me for the merger team and excluded Chris, which I think says a lot. He talks about empowerment, so I think I'm going to deal with this my way. It is not my problem that our software development is so slow. If I promise improvements to the customer, the business will have to play catch-up.'

Madison opened her mouth to challenge Tracy but changed her mind, as Tracy was already making moves to leave.

'Sure, Tracy, see you around.'

# THE BATTLE

Thomas grabbed two menus from the centre of the table and handed one to Gabriella.

'So Gabby, how are you enjoying your new role as CTO?' Thomas asked. 'This new lot wearing you down yet?'

'Wearing me down? Why? I'm having fun. We finally have funding for my ideas, and the leaders seem genuinely open to the strategic shift. Remember we merged with this company because they established themselves as industry leaders in designing and manufacturing medical test equipment. The designs are pretty strong from what I can see, with their products manufactured to high standards. What they are missing is the integration with personal wearable devices. They need a complete shift in their thinking, and that's where we come in. That was the whole point of the merger. Let's face it, we weren't having much fun at the old place with no funding and besides, you also got a promotion out of it. Director of Engineering. Go Tom!'

'Well, I'm not having much fun with Chris, the Engineering Manager. We've been working together for a month now and we seem to be constantly at loggerheads. He tries to stop or slow down everything I do. Chris is totally focused on existing products and current customer issues. Whenever I try to bring him into discussions about the shift in direction, he gives me these binary options. *Do you want me to ship the product, or do you want me to talk about future products?*'

'Don't get me wrong, even though Chris is a nightmare to manage, he's a talented engineer and an excellent manager. His team like him, bar Tracy, who seems to have it in for him. Chris does a good job. Chris is being coached by this business coach, Sonya, but I can't see it's making any difference. She must have told him to be more cooperative, as Chris walks around asking if he can help me. I don't need help. He needs help. What do I do?'

'I heard Chris likes squash. You're a sporty guy. Why don't you invite him out for a game at lunchtime? Do a bit of team building. I know you hate all this soft skill stuff, but we need this to work, Tom. *We* have the new ideas, but the existing products currently pay the bills.'

'I'm very unfit at the moment and haven't even got any squash kit. But hey, if playing squash is what it takes, that's what I'll do. Thanks, Gabriella. See you in the Ideation meeting.' He punctuated the word *ideation* in the air with speech marks.

'Hey, don't knock my initiatives. Creating a new language is all part of building a new culture. Thanks for the lunch. Now stop stressing out and go build your relationship with Chris.'

A few days later, Chris met Thomas at the squash courts during their lunch break.

'Ready to get slaughtered?' Thomas asked.

Chris watched Thomas strut onto the court. *Who does he think he is with his brand new outfit? I thought this was a friendly lunch break game?*

Chris felt glum. The sweaty smell of the squash courts reminded him of the locker rooms from his childhood football matches. His new boss triggered memories from those days - the bullying, the fear, the humiliation. *Lucky Thomas, the golden boy.*

'Let's do it,' Chris said as he grabbed the ball to serve. He slammed the ball against the wall and, from that moment on, continued to chase Thomas all over the court. Short shot. Long shot. Short shot. Left, right, left.

*Run, you bugger. Run.*

Thomas leaned against the glass wall, breathing heavily.

Chris looked at him with concern but was overcome by his memories of being bullied by the popular jocks at school and became determined again to show Thomas he wasn't a pushover.

Chris disguised a quick drop shot. Thomas sprinted to the front but completely missed the ball.

Chris glanced at him and immediately turned pale.

# FOLLOW MY LEADER

## LESSON 3 - IF YOU WANT TO LEAD, FIRST LEARN TO FOLLOW

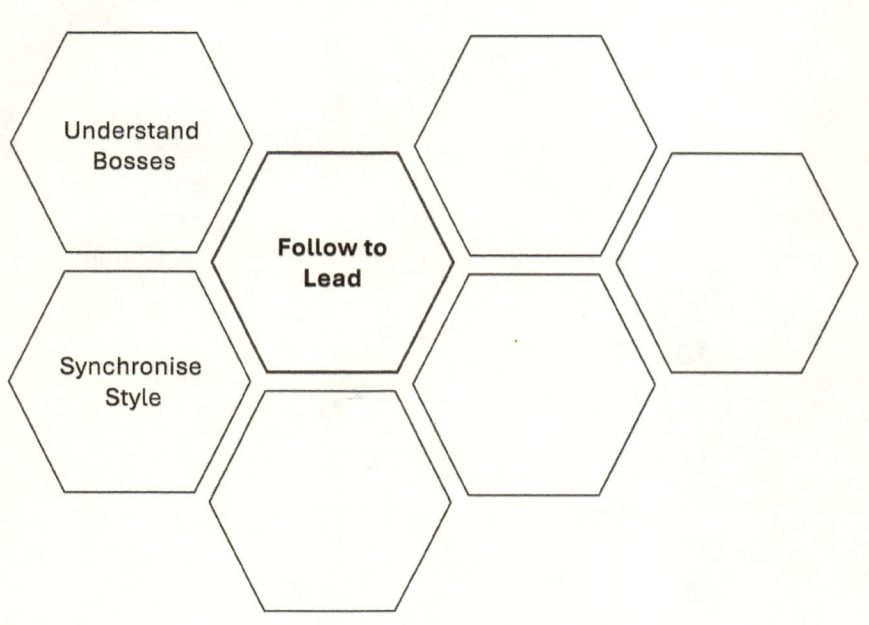

## HELP! I NEED TO MANAGE UP AND ACROSS

The news of Thomas's heart attack spread through the company like wildfire. Thomas wasn't even 50. Chris felt terrible about the whole thing but tried not to blame himself. *The squash match may well have been the catalyst. The guy was too driven, and probably the stress of the new job and having no clue about design pushed him over the edge. He'll get better.* Chris stopped at the stationary store to pick up a get-well card. He genuinely hoped Thomas would recover soon.

Meanwhile, the company appointed Gabriella, the new CTO, as acting Director of Engineering.

Chris called Sonya to schedule a time and place for their regular catch-up later that week.

'Sonya, I'm not sure what the benefit of this coaching is. They didn't even consider me for the *acting* Engineering Director role. Gabriella is now head of Engineering on top of all her responsibilities as CTO.'

'Chris, what are you doing differently now than when they appointed Thomas in the role?'

Chris paused, hoping she was asking a rhetorical question. Sonya remained silent.

'Sonya, maybe I can get more involved in this new stuff in a few months' time. Right now, I am up to my eyeballs in the current issues. This merger has doubled my workload, and they don't want to invest in the current product line as all the money has gone to Gabriella in R&D.'

'Chris, I can only imagine how tough things are for you.

Think of it from your boss' perspective. Why would they appoint you to *lead* the change if you cannot find time to understand what needs to change? There will never come a day when there are no current crises. You have to balance the short and long term. Gabriella is a temporary appointment, and you might find working with her easier than Thomas. It is a chance to get a fresh start on applying the Managing Up lessons you are learning. Who knows if Thomas will return to work once he gets out of the hospital? This may be an opportunity for you to show them what you've got.'

'Well, I am meeting with my new boss, Gabriella, today. I'll tell you how it goes when we meet up.'

After lunch, Chris met up with Gabriella. She had worked with Thomas in the previous company, so Chris thought he would start by asking her how he was doing.

'Tom is doing fine'. She was the only person who called him Tom. 'But he needs lots of rest. They are keeping him in hospital mainly as a precaution. In the lead-up to this merger, he and I worked 80-hour weeks, and Tom hadn't had a holiday in years. You've probably noticed how driven the guy is. This forced break will do him wonders.'

'Well, I hope he has a speedy recovery. I really am sorry about what happened.' Chris said.

Gabriella looked impassive. So Chris waited for her to make the next move.

'As you know, I will stand in for Tom until he is well

enough to return. I am sorry, I don't actually have the time to spend with you now. Rob has given me a lot of things to look into, and I am back-to-back with meetings. I need an analysis of the engineering design data from you. I've sent you a template to use with some key performance indices I am tracking. At the latest, I need that analysis by the end of play today. I will also need your design recommendations as an input to our Integration project.'

'Sure,' said Chris. 'You'll have the design analysis before you go home, and I'll send you my design recommendations for the Integration project by Friday.' He smiled at her, desperately trying to restore a bit of rapport again.

Chris returned to his desk and called his colleague, Dave, who was head of the Projects department and also now reported to Gabriella.

'Hi, Dave. How are you doing?'

'Probably better than you, buddy. I didn't put my new boss in the hospital.'

'Don't remind me. I feel sick about it. By the way, how are you getting on with Gabriella?'

'I have no idea. She's so busy I haven't caught up with her yet.'

'Lucky you. I just had 5 minutes with her, and she wants an analysis of all the design work we are doing. Can you believe it? It will take me all day. My biggest challenge is that I need additional design engineers, and without them, I end up doing the design work. Why is management not taking the initiative to address this problem? Instead, they're burdening me with unnecessary analysis.'

'Well, good luck mate.'

Chris felt his phone vibrate. 'Hey, I've got another call coming in. See you later.'

◆ ◆ ◆

'Quenton, hi.'

'Hi, Chris, Madison has asked me to sign off on the new contract to re-sign Beta Engineering. I specifically requested that the new contract align with our revised licensing policy, but she has gone ahead and proposed extending the existing unlimited user model.'

'Yes, I know. I support her approach. Madison increased pricing to current levels and besides, Beta Engineering will not be happy if we stop their existing licensing model.'

'Well, of course Beta Engineering will not be happy. This deal has been unprofitable for years. We have been waiting for the contract renewal to fix the commercials, and now Madison is throwing that opportunity away.'

'Quenton, you don't understand. If we move to the new model, we would have to upgrade the licenses for each existing user. It is a huge technical challenge. Madison has made all our lives easier, and I support her.'

'Well Chris, you're not the one who has to stand in front of our investment board and justify why our profits are falling. I cannot approve this renewal agreement. Your team needs to align this customer with our policy.'

'Whatever happened to putting the customer first?' Chris asked.

'Chris, this *is* putting the customer first. With our recent merger, it is essential that we get our licensing model consistent for all our platforms. This customer will not thank us in the long run if we make bad financial decisions, which means we can't invest in new products.

'Please tell Madison the renewal proposal is not approved and get her to renegotiate it.' Quenton put the phone down.

Chris called Madison and updated her.

'But Chris, I told you Quenton didn't like my suggestion, and you said you would convince him. Based on your approval, I have already verbally told the customer the deal was agreed to.'

'Well, unfortunately, Quenton is the CFO and he has the authority to stop the renewal. Sorry Madison, you are going to have to go back to the customer and renegotiate the deal.'

Madison put the phone down and groaned. Next time, I will make 100% sure I understand who needs to approve these deals before committing to the customer. Lesson learnt.

Madison picked up the phone again to call her contact at Beta Engineering. When she finished she said out loud, 'Well, that was fun. Not.'

Madison had accepted the rejection by Quenton, the finance director, of her Option A licensing model and had skillfully proposed Option B to the customer. This meant a conversion of all existing licensing to the new per-user licensing model on all their products. Her primary argument with the customer was that the alternative model was short-term and would ultimately damage their relationship. She pointed out that a financially unsound solution would harm them in the long run. It turned out that the customer had recognised that they were breaking the spirit of the original agreement. While criticising her for the turnaround in her proposal, they begrudgingly accepted the revised proposal.

At one level, she was angry with Chris for getting her into this mess with the customer, and at another, she blamed herself. *I could have been better prepared at the time. It is obvious now that Option A only had short-term benefits for us and actually would harm the customer long-term.*

She worked diligently at putting together a plan to convert all the customers' licences to *per-user* as agreed on the call. *When I think about it, thank goodness I am taking my pain pill now. Imagine doing this in a few years.*

Chris dropped in to see Fred, his veteran team member, who had asked to see him.

'Hi Fred, what's up?'

'Thanks for stopping by, Chris. I wanted to talk to you about the email Gabriella sent out this morning. When we were having the merger conversations, I explained our design and manufacturing process to Gabriella. She seems to have completely ignored that in her request. Gabriella is talking about a novel approach that ignores all our current processes. It is going to make our lives a lot harder. Can you talk to her and convince her to leave things the way they are?'

Chris reflected on what Sonya would say and then replied.

'Fred. One thing I can guarantee you is that things will not stay the same.

'Embrace the change and find out how you can influence it.'

Fred looked at him quizzically. 'Are you feeling alright? You're happy with the change?'

'Fred, it is not about being happy with the changes. To be honest, I was struggling with this as well and fighting against it. My business coach, Sonya, gave me the insight that to influence the change, I need to be part of it. If I am negative and seen as resisting change, I will be excluded and have no way to influence it positively. She also highlighted to me that to progress, change is inevitable. Our challenge is to ensure the inevitable changes do indeed improve the outcome. Embrace Gabriella's fresh approach and provide constructive feedback on what you like about it. Ask her good questions that will help to delve into your concerns. She will sideline you if she perceives you as being against her ideas. Make sure you are part of the solution.'

'That sounds like hard work to me. I'm not sure I have the energy. I enjoyed being on that merger team where people listened to my input without questioning it. Now that we're back to operational mode I find it boring again.'

'Just waiting things out till your retirement, are you, Fred?' Chris mused. 'That might work for you, but I'm not sure it works for me. Sorry, I need to run. Please think about what I said.'

Chris walked back to his office, deep in thought. *Watching people like Fred gives me good insights into how I may look to my managers. If I want others to follow my leadership, I need to get better at following, too.*

# RIGHT OUTCOMES DESPITE MY BOSS

## LESSON 4 - USE THE TWO-UP-TWO-ACROSS RULE TO PROBE FOR THE PURPOSE OF THE REQUEST.

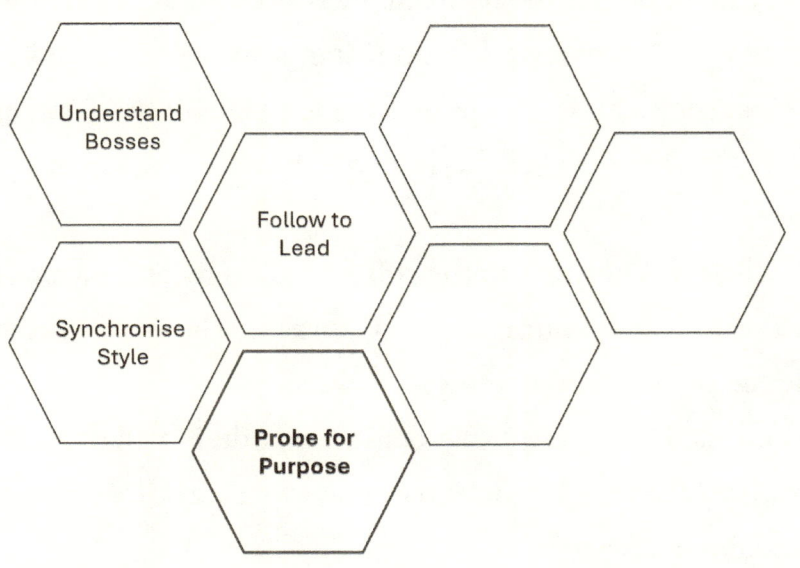

The following morning, Gabriella rushed up to the water cooler to fill her water bottle as she ran between meetings.

Rob walked past and asked how she was.

'I'm doing fine, Rob. I don't have to deal with all you have as a new CEO, so I can't complain.'

Rob replied, 'I can't believe what happened to Thomas, but I am glad he is recovering. By the way, please don't forget the engineering design data I need. I have a hunch that we need to increase the number of people in the design team, but I need the data to prove it. The meeting is this week, and I want the data to include in my presentation to the board to approve additional design hires.'

'I haven't forgotten, Rob. I reminded Chris again when we last met. I have no idea why he is taking so long with it. I'll chase him again today.'

'And please use the template I gave you. It is part of a bigger analytical framework I am using to make our decision.'

'No problem, Rob. I passed the template onto Chris, and I'll make sure you get the info you need in the format you want.'

Chris was heading to the coffee shop to meet Sonya when he almost literally bumped into Gabriella, who was seemingly late for yet another meeting.

'Chris, I really need that engineering design analysis. You promised it to me by the end of play yesterday. Rob is on my back about it.'

'I emailed it to you earlier. It's in your inbox. I knew it was important to you, so I worked on it last night at home.'

Gabriella checked her phone and scrolled through her emails. 'Chris, this report is five pages long, and I can't read the figures on your spreadsheet. Can you please send me a summary of the key data and metrics I requested, using the template I provided you? Please send it before midday. I've got to run.'

Chris continued over to the coffee shop. 'My new boss is driving me mad,' Chris said to Sonya as he sat down.

'But I heard he was still in hospital. What has he done now?'

'Not Thomas. His replacement, Gabriella. Yesterday, she asked me for an analysis of the design work I was doing. I worked on it all day and even had to work on it after hours, trying to be a good follower. Today, she ignored the report, saying she didn't have time to read it, and asked me to complete her template, which is a bunch of Key Performance Metrics and data. I desperately need extra staff and she is playing around analysing numbers and metrics.'

'Chris, one lesson in Managing Up is that your bosses seldom ask you for what they actually need. What they really need originates at the next level up or across the organisation, and not all bosses have the expertise or communication skills to translate that into exactly what is required for you to do. You first need to find out the purpose of the request - the why. It's what I call the *two-up-two-across rule*. If you want to know what your boss really wants, first probe for purpose. Find out who - up or across the organisation - asked for it, and then identify what *they* really wanted.'

'But how can I do that? I can hardly say, *Gabriella, I don't think you really know what you are asking for. Why do you want this?*'

'You are right. *Why* is a trigger word. You need to find words that get you to understand the *reason* for the request, without using the word *why*. Something like, "So that I can get you exactly what you need, can I understand what you plan to do with the information I am sending you?" Gabriella doesn't want that information for herself. She wants it to give to someone else. If you can find out what *they* really want, you will understand the true request.'

'Thanks, Sonya, that's really helpful. I suppose it's the same as dealing with customers. They say, *the customer is always right*, but the customer is rarely right. How can they be? They don't understand the details of what we do, so what they ask for is often not possible or practical. We need to ask good questions to find out what they really need.'

'You've nailed it, Chris,' said Sonya as she gathered her things to leave.

Chris returned to the office in much better spirits than before his meeting with Sonya. He wanted to work on Gabriella's request, amongst a few other priorities, but first, he wanted to understand her real request, so he called her using his new language.

'Hi Gabriella. So I can give you exactly what you need, can I understand what you are going to do with this information?'

As Gabriella spoke, Chris realised he had completely misunderstood the request. *Oh — for Rob — key metrics for the board — hoping to use it to get more design staff.*

'Sorry, Gabriella. I completely misunderstood your request. I'll make it a top priority and send it by midday, using the template you provided.'

After an hour of working on the key metrics and ensuring it clearly highlighted their resourcing problem, he sent it to Gabriella.

The next day, Gabriella dropped in to see Chris in his office. 'How is everything going?' she asked.

'All good,' said Chris. 'You?'

'I'm good. Thanks for the design analysis you sent for Rob. He seemed happy with it. Hopefully, we can get you an increased budget for additional staff now. Regarding those design recommendations I will need them before Friday.'

'I had planned to do that by the end of the day on Friday. I have a long list of actions from the executive management meeting, and Chuck keeps dragging me into urgent customer issues. Can we stick to the original deadline?'

'Chris, that won't work for me. Your design input is critical to the new product release and I have already committed to Rob to release it this month. I'll give you another 24 hours, then I need it, please.'

Gabriella stood up and left Chris's office without looking back at him.

Chris was feeling too stressed to do any work. It was just before his lunch break, so he grabbed his packed lunch with his Management book and headed off to a local park to de-stress and get help with his conundrum.

The park was virtually deserted except for a few mothers with strollers chatting away while watching the older toddlers on the playground equipment. The leaves rustled rhythmically in the trees, accompanied by the odd bird call. The air smelt fresh and fragrant. The sun shone down on the ground as it had for millions of years. Chris breathed deeply and suddenly felt calmer. He started reading.

## CONFLICTING PRIORITIES

While it's true that multiple virtual bosses exist and the line manager may not always be the most influential, it's crucial not to disregard their directives. The line manager is uniquely responsible for enforcing minimum compliance. Their authority comes from their role, not necessarily any special skills or personal influence. In the future, it is possible that the supervisory line management component could be done by AI tools, as the compliance line is completely objective.

Line managers have authority that comes from the

organisational structure, not the person. Senior managers outside the organisational hierarchy do not have the authority to override a line manager who is executing their organisational authority. These managers have to use their influence skills to prioritise long-term issues. Understanding the authority and business impact of requests from bosses outside the hierarchy is critical to being effective in managing up.

While valuable contributions to the future may extend beyond core job responsibilities, managers must ensure their mandated duties remain the top priority. This strategic approach ensures that essential tasks are addressed first, before allocating resources to supplementary activities. By doing so, managers can feel focused and efficient, knowing they are managing their time and resources effectively.

Chris took the last bite of his sandwich and closed the book. Looks like letting Gabriella down because I am working on important activities that her boss, Rob, has given me is not the best idea. Interesting.

The next day, Chris had his regular weekly catchup with Sonya and addressed his other concerns.

'Sonya, I found the insights on the multiple bosses interesting. I do have questions. What do I do if my boss changes my priorities, which stops me from doing something else that benefits the business? I find my bosses do not always understand the importance of the existing work I am doing.'

'Chris, it is always in your boss's interest for you to succeed. Everything you do is a reflection of them. If they have asked you to do something that will indirectly damage the business, it is part of your responsibility to spell out the consequences so they understand the impact of their request. They will not have knowingly asked you to do something that will damage the business.

'Remember also that *you* may have misread the relative importance of the changed priority. In the knowledge era, the *worker* often understands the impact of task-oriented work more clearly than the boss, but the boss may have a clearer understanding of the bigger company picture. Communication on changed priorities needs to be two-way.'

After his meeting with Sonya, Chris went back to his office. He picked up the phone to Chuck. 'Yes, I know I said I would attend the meeting. Fred knows the technical issues as well as I do and I have to urgently complete some critical input to Gabriella's Integration project.'

'Since when do internal issues come before customers?' asked Chuck.

'Chuck. Vitality Inc. only has 4 licences. They are one of our smallest customers. The work I am doing for Gabriella affects all our customers, large and small. Fred is ready and prepared to attend.'

Chris ended his call with Chuck and prioritised completing the design recommendations for Gabriella.

A little later he received a call from Tracy. She explained that her dog was very sick, and she needed to take him to the vet the following morning. She didn't want to put a leave form in, as she felt she could work from the waiting room and in practise would only be off work for the half hour consult itself. She also wanted to work from home for the remainder of the day to monitor her dog and make sure he was okay.

Chris, who also had two dogs, empathised with Tracy's situation. 'No problem, Tracy. I understand. I hope everything goes well and your dog recovers soon. It's always a worry when our pets are sick. They become part of the family.'

Tracy thanked Chris and then phoned a friend.

'All sorted. I can join you at the spa day tomorrow. My boss is such a pushover. I told him I had to take my dog to the vet. I don't even have a dog.'

# THE SALES MEETING

A flip chart with fresh paper stood at the front of the room, its steel legs protruding awkwardly like a giraffe. The windowless room echoed with people scraping chairs and setting up their laptops on the long steel table in the middle. Charger cables obstructed walkways to refresh mobile phones, laptops and headsets. A large white screen dangled precariously from a cable secured to the wall facing an overhead projector bolted to the ceiling.

Chris found an empty seat and sat down. *Looks like budget cuts have kicked in. No 5-star convention centre like last year.*

Chuck rose slowly and walked to the front of the room. 'Okay, we are all here. Let's get started.'

Chris stopped him and pointed out one or two salespeople were still missing.

'Well, they won't work here long if they are late, so don't worry about them. Let's get going.'

Chris nodded and looked around the room, silently confirming that all of his presales engineers were present.

Chuck set up a Bluetooth connection from his laptop to the projector and played a movie clip from *Wolf of Wall Street*.

*'So you listen to me, and you listen well. Are you behind on your credit card bills? Good, pick up the phone and start dialling! Is your landlord ready to evict you? Good! Pick up the phone and start dialling! Does your girlfriend think you're a worthless loser? Good! Pick up the phone and start dialling! I want you to deal with your problems by becoming rich!'*

Chuck hit stop on the video clip. He stood up and, without speaking, looked around the room slowly.

'Ladies and gentlemen. Welcome to the world of sales. Sales is for winners, not whiners. Sales is for people with guts, not wimps. Sales is for people who hit targets, not make excuses. Is there anyone here who doesn't want to be a winner? Please leave.' Chuck waited and looked around.

Chris felt like he was at a wedding and waiting for any objections when the priest asks if anybody objects to the union of two people.

After what felt like an eternity, Chuck spoke again. 'Right, so let me ask you a question. If you are all winners, why are we behind on our numbers?'

Chuck returned to his laptop and clicked on a button.

The screen filled up with names, quotas, and % achievement figures. A bright red line divided those on target and those below target.

All the salespeople, including those above the line, looked at their feet. Chris shuffled uncomfortably in his chair. Not all of his team members looked affected by the drama of the moment. Fred was tapping away on his laptop, but some of the new people who had joined with the merger looked at each other with wide eyes.

'And you presales engineering folk. What are *you* doing to help us hit the target? I don't care about your geeky designs. Our customers don't care about your geeky designs. They want *value*. And if you were delivering value, we would be on target. We need to lift our game, people. Lift our game.'

One brave soul raised an objection. 'Chuck, with due respect, some of these accounts we are working on are strategic and long-term. We can't rush the customer.'

'I will let you into a little secret. If this team doesn't get back on target, there will be no long-term. Listen, if you can't handle the heat, get out of the kitchen.'

With that, Chuck grabbed a marker pen, walked over to the flip-chart, and dramatically turned over a clean sheet of butcher paper.

'Okay, we will go through this forecast account-by-account and line-by-line, and you better be prepared. Let's start with you, Ronald. Who are the presales engineers helping you close the Medfast account? Let's hear how you plan to get this account back on track.'

Chris groaned inwardly but maintained a poker face as Chuck interrogated each team member. *Rob would hate this. I need to stop him.* Chris fumbled for his phone and tried to look innocent as he pressed *record* on his mobile phone.

Chuck was in the middle of another rant when his mobile went off. 'This is important,' he said as he left the room.

After a few minutes, Chuck walked back in and said. 'Something important has come up and I need to go over to the client immediately. Chris, could you continue going through the account plans with the team? I will be back later. And please do not interrupt me by calling while I am gone. I need to save this account.'

Once Chuck had left the room, Chris used the opportunity to give everyone a 20-minute break and invited them all over to the coffee shop for a morning pick-me-up. Once they had ordered, he left them to drink their coffees and enjoy the fresh doughnuts. He returned to the room to prepare for the rest of the meeting. He realised that if he was to be presenting, his laptop power wouldn't last, and he had left his charger in his car.

Chris headed down to the underground parking lot to fetch the charger. He noticed a big black BMW with dark tinted windows, parked next to Chuck's fancy sports car - *how does such a big guy fit into that tiny car?*

Someone inside the BMW lit up a cigarette, silhouetting two figures through the glass. One of them was a rather large man. *Chuck?*

Chris looked away quickly and busied himself, retrieving his laptop charger from his car. *Could that have been Chuck? Why was his car still in the parking lot? What is going on?*

Chris put the mystery of Chuck and the black BMW behind him. With everyone in good spirits from their visit to the coffee shop, Chris changed tack for the meeting. He put his laptop, with its blue charging light blinking furiously to the side, and addressed the team.

'I'm guessing if anyone cares about getting these accounts back on track, it is you, considering half your compensation is riding on meeting these sales targets. So, let's assume that a lack of motivation or refusal to work hard is not the problem. Who wants to tell me what the *real* issue is?'

One of the more experienced members of the sales team spoke up. 'Chuck set these targets as stretch targets with no correlation to the opportunities we are working on. Many of these accounts are complex, and to win long-term business, we need to build relationships and understand the customers' real needs.'

Someone else chimed in, 'Pushing our product down their throat with aggressive pricing may meet our short-term targets, but it's at the expense of long-term relationships and ultimately puts the growth of our accounts at risk.'

Chris listened carefully as each salesperson discussed what they believed would really help to progress their account plans.

The most experienced salesperson in the room concluded the discussion by saying, 'Chris, we all understand the need to meet short-term targets. Heck, half our remuneration is tied up in meeting those quotas. All we are asking is that we balance them with long-term growth.'

Chris wrote up all the actions the team had committed to and promised to discuss the sales team's concerns with Chuck. With that, he closed the meeting.

Later in the day, he walked past Chuck's office and dropped in for a chat. He opened his laptop, showed Chuck the action plans he had documented, and then gingerly addressed the team's concerns about targets.

'Chris, thanks for helping out today, but I don't interfere with your engineering designs, so please don't come in here and lecture me on how to run my sales team. You are too soft, and those guys were playing you. If anything, these targets are too low. Next year, I will be more aggressive, and those who can't handle the pressure need to get another job.'

Chris changed the topic. 'How did your customer meeting go today?'

'Huh? What meeting?'

'You know, the customer you left the sales meeting for?'

'Oh yeah, of course. It went so well that I forgot all about it. It's all good buddy, all good. Listen, sorry to rush you out of my office, but I have calls to make. Close the door behind you, won't you?'

Chris decided to pop in and see his old boss Rob to see if he could get support for the revised sales targets for his presales team and fight the battle for Chuck's sales team. Rob was the CEO now, and Chris had engaged in enough conversations over the years to know Rob would not like what had happened in the sales meeting today.

Rob listened carefully.

'If you like, I can show you what happened today. I recorded some of it.' Chris took his phone out, but Rob stopped him.

'Chris. Put your phone away. I know you have a lot on your plate in your new role. Chuck can come across as aggressive sometimes, but he knows what he is doing. Why don't you focus on what you need to do in engineering and let Chuck get on with making the numbers?'

Chris walked out of Rob's office as Quenton was arriving. *I guess I've been told. So much for managing up and across and trying to help. Perhaps I should wind my neck in and mind my own business.*

Quenton sat down to talk to Rob.

'Rob, I wanted to run something by you. Chuck called me earlier, saying he urgently needed a copy of our internal costing sheet for a deal he was negotiating. As you know, we have pretty confidential company data in that pack and we rarely share that with the sales team. He explained to me how urgent it was and said that you would be fine with it and you knew all about the deal he was working on. He didn't want to disturb you and promised me he would keep the information confidential. I gave him the data, but now I am nervous, I should have checked with you first.'

'Thanks, Quenton. I appreciate the heads-up. We can't micromanage our sales team, and if I can't trust Chuck, then who can I trust?'

'Good point, Rob. Okay, thanks, have a good evening.'

The following morning, Tracy met Madison, her co-worker, for breakfast. They both ordered the *chef's special*, along with an oat milk latte. The breakfast arrived with avocado slices beautifully laid out on freshly baked sourdough bread with cracked black pepper and fresh thyme leaves sprinkled on top. A few slices of smoked salmon were included, garnished with cherry tomatoes.

'I love this place,' said Tracy. 'I wish I could say the same about work.'

'What's wrong?' said Madison.

'Just the usual. Nothing changes. Chris asked me to take on another project, *Autotest*. I am *already* snowed under. I think I handled it well. I told him I could only handle two projects. As he was the boss he could choose which two.'

'Wow. What did Chris say?'

'What could he say? He realised he was over-stretching me and dropped it.'

'Well, that's great for *you,* Tracy. I can imagine he will ask one of us now. We are also busy, you know.'

'Not my problem, Madison. I can't help it if people don't stand up to Chris. Chris really should take on some projects himself. He doesn't seem to do much, other than sit in meetings all day.'

'Okay, Tracy. Let's not get personal. Hey, how was that concert you went to on Monday night?'

'Yeah, fine, let's order another coffee, and I will tell you all about it.'

'Let's take a rain check on the coffees. We are already late for work, and I have a ton of work to do. Tell me about it as we walk back to the office.'

'Madison, you work too hard. Nobody is going to miss us, and besides, I don't feel guilty about being late. I put lots of extra hours in. Last Monday, I left the office half an hour after knock-off time. I don't even know why we are talking about it. This is what I hate about this company. It's so old-fashioned. Always watching the clock.'

'Tracy, for me, it's not about watching the clock. Nobody is forcing me back to the office. I have a lot to do, and I want to get on with it.'

'Sure, Madison, you go back. I'm going to order another

coffee and make some calls from here. If you're not interested in hearing about the concert, I know someone who is.'

Madison departed, placing some cash on the table for her half of the meal.

# MAP YOUR ALLIES

## LESSON 5: NETWORK WITH POWER BROKERS TO INCREASE YOUR INFLUENCE

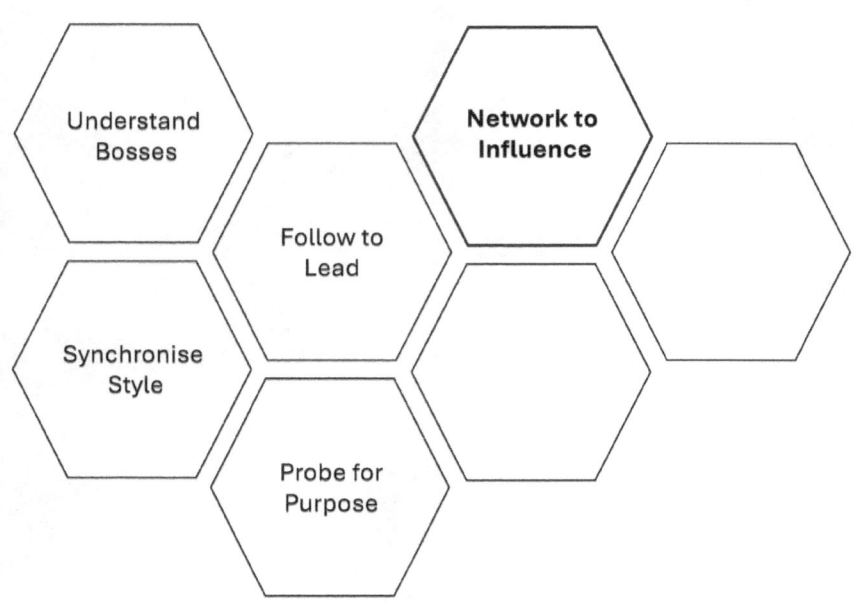

Chris, visibly perplexed, sat down with Sonya and cradled his coffee, his gaze fixed on it as he spoke. 'Sonya, I'm at a loss here. I did what I thought was right by approaching Chuck first, and when he brushed me off, I turned to Rob. I've known Rob for years. I know he values maintaining long-term customer relationships over arbitrary targets. So why did Rob dismiss me like that?'

Sonya replied. 'Chris, who does Chuck report to?'

'He reports to Rob.' Chris said.

'And so who keeps Chuck hired and who approves the sales targets?'

'Well, yes, Rob.'

'So when you criticise Chuck and the targets, who are you really criticising?'

Chris stared deeper into his coffee cup. 'I suppose I am criticising Rob. But then, what was I supposed to do? I thought you said managing up and across the business was important. Minding my own business and ignoring the issue hardly seems like being a team player. My problem is I have no power. Maybe if I could get Rob to combine the sales and engineering teams and put me in charge, I could do something about it.'

'Chris, you are right. You don't have any authority over people who don't report to you. You can't use your title and rank to insist that things get done by punishing or rewarding them for complying with your request. But there are several other sources of power.

'Your expertise is a source of power, especially if it is relevant and scarce.'

'You're right, Sonya. They say knowledge is power, although these days, with AI beating humans at their own game, I'm not sure it is as important as it used to be.'

'Yeah, good point. Your status and prestige are also a form of power because your bosses get recognition based on your reputation.'

'I agree with that. Management respects the most influential people in my team.'

'Don't underestimate your personal power based on your personality.'

'You mean what's called referent power?'

'Yes, exactly. Warmth, optimism, enthusiasm, humility, assertiveness and trustworthiness will go a long way towards influencing others. You also need to learn lessons in networking with key players who have influence.'

'Networking? Networking and hob nobbing will not help. I don't want to play power games. I have proof of his behaviour. I recorded part of the meeting and—'

'You recorded it? You know it is illegal to record someone without their knowledge in this state? I bet you it is against company policy, too. No wonder Rob cut you off. He was trying to *protect* you.'

'Well, isn't bullying also against company policy? I am trying to protect the staff from Chuck's unacceptable behaviour.'

'Chris, your motives are good, but I think you are going about it the wrong way. I suggest you read up on the chapter on networking and influence, and let's discuss it in our session next week.'

Later that day, Chris sat down to read.

## NETWORKING AND INFLUENCE

In the workplace, executives want to achieve good results and hope to achieve them through a known strategy. If they can figure out the cause-effect of what makes the company grow, they can invest in those levers of success.

At the company's nucleus is a core team of change champions who uphold this strategic model. These players are the chief influencers who shape and direct the strategy. Forming an alliance with these power brokers can increase one's influence.

A key lesson in networking is to establish strategic allies *before* you need them.

Networking is a proactive activity which involves identifying and engaging with key stakeholders who hold sway over critical resources and budget allocations. Networking is not a self-gratuitous activity to garner personal favours or popularity; it embodies a reciprocal exchange predicated on mutual benefit. The key to influencing others is to start by understanding the needs and aspirations of the other person. By discerning *their* objectives and motivations, individuals can tailor their networking efforts to align with the interests of these influential decision-makers, thereby enhancing their capacity to exert influence and effect change.

The organisational chart represents a formal structure developed from historical insights. It delineates lines of authority and clusters job tasks into logical groupings, providing a clear framework for operational efficiency.

For the activities that contribute to the business growth, numerous uncertainties make it challenging to immediately integrate these new initiatives into the established structure. Instead, these initiatives often operate in an informal "start-up mode" alongside the formal organisation. To lead effectively in this context, understanding the informal hierarchy is crucial. This includes identifying and cultivating relationships with key players who drive influence within this dynamic structure.

One practical approach is to develop a stakeholder map that highlights the future-orientated "movers and shakers." Strategic engagement with these individuals can help align initiatives with broader goals, allowing leaders to enhance their influence and drive growth within the informal network.

◆ ◆ ◆

Chris slammed the book shut. He knew exactly what to say when he met up with Sonya, which happened the following Friday.

'Did you read the section on influence?' Sonya asked.

'Yes, I did,' Chris replied smugly.

'Oh good.' Sonya replied. 'Have you heard of the six degrees of separation and how it relates to Kevin Bacon?'

'No?'

Sonya enlightened him. 'The six degrees of separation is the idea that everyone is connected through a chain of six or fewer social connections. In other words, you could be just a few introductions away from anyone on the planet. Kevin Bacon got famously tied to this concept with a fun game called Six Degrees of Kevin Bacon, where people try to link any actor to him in six steps or less based on movie roles they've shared. It turned into a whole pop culture phenomenon because he's been in so many films, making it surprisingly easy to link him to almost any actor. Kevin Bacon was the super-connector. If you want to increase your influence in the company or industry, build a relationship with super connectors.'

Chris smiled gratefully.

Sonya continued. 'Earlier, you looked like you had something to say, regarding the book segment you read?'

'Yes, I did. I realised the mistake I made when I visited Rob was that it was all about me. Whenever I see Rob, he asks me how I am and what he can do to help. I seldom give a second thought to what he needs.'

'Okay, Chris, let's work through this. So you want Chuck to change his behaviour towards the sales team. You think it is hurting morale. Rob cares about morale, too, but Rob also cares about meeting the numbers. I assume Rob also has had no formal complaints about Chuck's behaviour.'

'That's because they are too scared to report him.'

'Chris, I get it, but look at it from Rob's perspective. He has not witnessed Chuck's behaviour personally and knows he cannot watch your *evidence* filmed secretly. You are trying to put Rob in a position where he needs to pick sides. The way you have presented the options, if Rob chooses you, he won't get his sales numbers. Go away and think about what wins you can give Rob; that will also give you a win.'

'Got it, Sonya. No more all-or-nothing thinking. You've said before that the best way to influence someone is to help them achieve what *they* want. I'll change my approach and get back to you on how it works out.'

Back in his office, Chris heard that a key customer was rejecting the new architectural design, using the merged company's new strategy of integrating all new products with wearables. The customer argued that they had no plans to incorporate wearables in their company right now. Chris recalled a LinkedIn post by one of their company's executives, who had written a thought leadership piece on the topic of wearables. Chris had sent a connection request, which the executive accepted. They exchanged a few positive comments on the article and traded messages through LinkedIn.

Remembering that the executive's name was also Chris, he searched his LinkedIn contacts.

*Christopher Wilson. That's him.*

He reviewed Christopher's LinkedIn profile and his various posts to gain insights into the company's strategic plans. Armed with this information, Chris strengthened his case to convince the customer he was dealing with, that this solution had good long-term value. Alternatively, Chris reasoned, maybe Christopher could help to smooth the way for him, given his passion for the topic. *Clearly the person I am dealing with is a bit out of touch with the strategic direction of their company. Skilful networking could help me here.*

After a busy day Chris arrived home feeling tired and grumpy. Liam, his oldest son, walked into the lounge room. 'Hi, Dad, can I get you a drink? You look tired.'

'Thanks, buddy. There is a cold beer in the fridge. Help yourself to a soda while you are there.'

Liam came through the door carrying Chris's beer plus a small bowl of peanuts.

'You really are a lifesaver, young man. Tell me, how was your day today?'

'Yeah, good, but I'm upset about tomorrow.'

'Tomorrow? Why, what is happening tomorrow?'

'Well, I was hoping to play football, but Mom is working, and she said you had a golf tournament, so I can't get there.'

'That's too bad, buddy. Yes, I am sorry, but I am leaving very early for the tournament, so there is no way I can drop you off.'

'Hey, Dad. I have an idea. After dinner, could you drop me at Charlie's house? He is going tomorrow and his parents said it is fine for me to sleep over. He lives near the driving range, so you could get some final practice before your match tomorrow. Please?'

'I will check with your mother and if she is okay with it, so am I.'

'Dad, I already asked her, and she said I could go as long as you could drop me off tonight.'

Chris smiled at his son. He certainly had learnt the lessons of managing up.

'Sure, buddy. Let's go after dinner.'

# HOW TO SAY NO TO MY BOSS

**LESSON 6: LEARN THE 'YES AND' STRATEGY TO SAY NO, BY NEGOTIATING THE QUALITY LEVERS OF TIME, MONEY, SCOPE.**

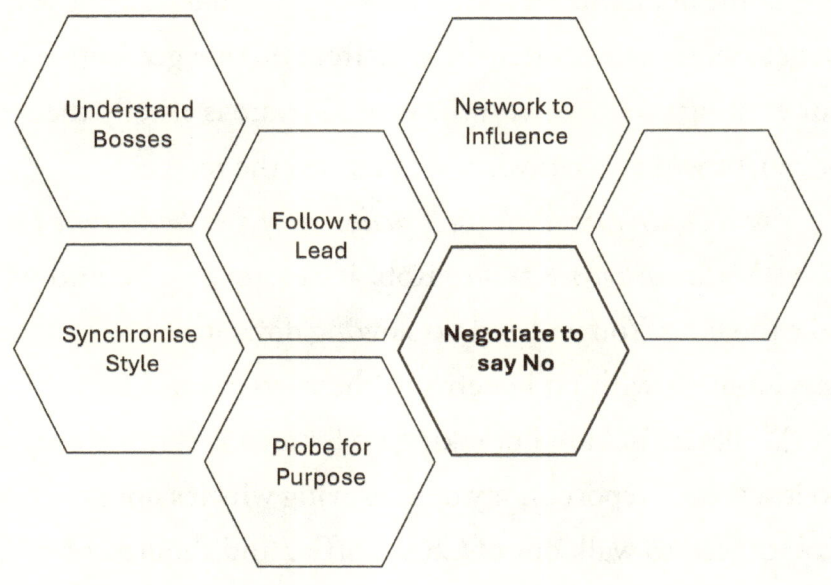

It was Wednesday morning, and Chris was sitting in his office, focused on the design recommendations for Gabriella's Integration project. Rob walked in and sat down on a visitor's chair.

'Chris, I have been thinking.'

Chris pressed save on his document and looked up. *Not good. More distractions to fend off.*

'Oh yeah? Thinking about what?'

'Well, what if we combined the work we are doing on the new Integration product with a code change to the old one? Gabriella tells me the new software has massive efficiencies and runs at four times the speed.'

'But Rob, we need additional resources to handle both projects. And besides, it's too complex to change the software of the old product now. It runs on a different architecture.'

'Come on, Chris. You are a smart guy. You'll figure it out. I want to see some short-term benefits from this merger. Improving our existing products with the new software is exactly the sort of benefit we expected when we planned the merger.'

*You wouldn't have all these delusions if you had invited me onto the due diligence team.* 'Rob, it's too expensive and will take too long. You will end up slowing down the new product development with no benefits to the current one.'

'Well, Chris. I am not taking no for an answer. If you can't do it, I want a report on my desk proving why it's not possible.' Rob turned to walk out of Chris's office and shouted over his shoulder. 'We need innovative thinking around here, Chris. Talk to Gabriella.'

Chris returned to his laptop. 'What's the point of finishing this report? Looks like it's all going to change anyway.'

Chris reviewed the project schedule and realised he also needed to deal with the new *Autotest* project that kicked off earlier in the week. *Tracy has not stepped up to the extent I expected, but I can't force her to take on extra tasks. Madison is better at managing her workload. I'll get her involved.* Chris called Madison in to see him.

'Somehow, I thought you may knock on my door,' Madison said with a smile.

'What?'

'No, nothing. Tell me about the customer requirements for this new *Autotest* project.'

'How did you know it was Autotest?' Chris said.

'Never mind. Tell me about it.'

Chris explained what was required.

Madison replied. 'I can see it is pretty important that we successfully deliver *Autotest*. Let me share with you what I already have on my plate so we can rearrange our priorities. We might have to make some tradeoffs on other projects to fit this in.'

They discussed their options, and then Chris asked, 'Are you absolutely sure we can't deliver *Autotest* without moving the dates of the other projects?'

'100%. I understand we have to deliver Autotest, and we don't have a budget for additional staff, so we need to renegotiate the date of the other projects. They are less strategic than this, so they must be the *victims*.'

'Okay, thanks, Madison. I'll leave it with you to get a team on *Autotest* and produce the plan. I appreciate your help.'

On Friday, Chris had his weekly catchup with Sonya, and they agreed to meet over coffee. It was 3pm. The coffee shop was relatively quiet, with half the chairs packed up, ready for closing. At this time of day, they served the coffee in takeaway cups, and Chris was unimpressed. 'I hate drinking good coffee in a paper cup,' he told Sonya.

'I disagree. The coffee is lovely and hot in a takeaway cup. In the future, I'll always order my coffee in a takeaway cup. Anyway, you mentioned you were struggling to say *no*. What's been making that difficult lately?'

Chris explained the situation and waited for Sonya to respond.

'You probably got to your senior role by saying yes. Your challenge now is that if you keep saying yes to everyone, in practice, you are saying no to someone else. Managers and customers have what I call the empty calendar delusion. Their primary assumption is that you are working on the right things and your time is optimised. However, when they make an additional request, they assume time and other resources magically appear, and expect you to respond without affecting your current tasks.

'You turned down Rob's request, quoting *your* limitations, not his. The fact that it is too expensive, takes too long, or is technically challenging may be worth it to Rob if he gets the win he wants. I guess Rob sees those things as *your* constraints, not his.'

Sonya continued, 'Chris, I am guessing you have a full calendar of things to do and, if anything, are wondering how you will get everything done. The problem with saying yes is you may not have any free time to do extra activities.

'Share the dilemma with the person by negotiating what needs to *not* be done. This will free up space to do the new task. From your boss's perspective, they hired you to do whatever needs to be done, so they expect a *"yes"* from you when they request something. Use the *"yes, and"* strategy. The *"yes"* part shows them that you will, in principle, do anything they ask you to do. They set the priorities, not you. The *"and"* part shares the consequences of saying yes and negotiates what needs to change to address the new request. What are you now NOT going to do, or certainly not going to do in the way originally planned — with reduced scope, or extended timelines, or a request for an increased budget?'

'Great insight, Sonya. You're right. I am mainly worrying about the lack of resources and the inevitable delay in current projects because it will affect the overall business. Rob needs to understand the impact on him, not me. I'd better use the same technique with Gabriella, or we will have one almighty battle on our hands.'

Sonya replied. 'In project management, the definition of a quality outcome is meeting the time, money and scope constraints. This is different to how people generally think of quality, where it is an absolute concept divorced from time, money and scope constraints. Having an absolute idea in your head what quality means is often why you say no to your boss. You don't believe you can do it within time, money and scope constraints. Instead of saying an outright no, you can find a way forward by negotiating the flexibility of time, money or scope levers.'

'Makes sense, Sonya. If Rob saw my consequences as *his*, he may have a different view. And if he doesn't, he will buy into my "no" decision. I'm meeting Gabriella on this topic, so I will use this insight with her.'

Chris left the coffee shop and headed to Gabriella's office, hoping she hadn't left for the weekend. When he entered, there was a strange smell of incense and whale music playing in the background. She had some odd sized, colourful crystals on her desk. She looked up slowly as though he had interrupted her meditation.

'What can I do for you, Chris?'

'Hi Gabriella. I want to discuss Rob's idea about integrating the new software into our existing products with you. It would benefit the company, so I want to see what could be done to make it happen.'

Gabriella looked at him, trying to see if he was being facetious. 'Seriously? You like the idea? I thought you would be dead against it. You have quite a reputation around here for pouring cold water on new ideas, Chris. I'm glad you are open to discussing this. So you think it's a go?'

'Gabriella, as a concept, I am really for it. My concerns are around its impact on other priorities and the budget. Putting that to one side for now, here are some things we need to do to make it work.' Chris sat forward in his chair and responded.

'First, we need to put the current Optimisation project on hold or at least slow it down. We will need all hands on deck to do this integration exercise.'

'But aren't we about to ship a major software revision to one of our largest customers?'

'Yes, Gabriella, we are. We can't do both, so we will need to balance the benefits of accelerating the new Integration project against the impact on our existing customers. I want to be sure my opinion on that matches the company's. Let's come back to that later. The second thing we need is increased resourcing. I made a quick estimate and identified some consultants in the market who could help us. They have done a similar project for another customer. Here's the rough budget approval I need.' Chris wrote a figure down and showed it to Gabriella.

'You're kidding, right? Where do you think I can get that sort of money?'

'Talk to Rob. He is dead keen on this. Rob might have a pot of money he can dedicate to this—'

'Chris, stop. I can guarantee you there is no pot of money for this. In fact, Rob is looking for short-term cost savings. Besides, Rob will certainly not upset our largest and most loyal customer. Leave it with me. Sorry, Chris. I think we need to delay this software integration idea and complete the optimisation project.'

'Thanks, Gabriella. I'm glad we are aligned on this. If I come across as negative sometimes, it's because I've decided that variables cannot be shifted. It is good to confirm those with you so we are both on the same page.'

'Absolutely. The reality is we can't do everything so there is always something that has to give. You're right. It does come across as negative when someone appears to have a closed mind and keeps saying something cannot be done rather than exploring what *can* be done.

'By the way, Chris, did you see we lost the Brenner account? Our competitor undercut our pricing. They also had critical information about our roadmap, including delays in our supply chain. How on earth would they know that? Could anyone on your team be leaking information? Chuck is furious about it.'

# ESCALATING THE CRISIS

## LESSON 7 - COMMUNICATE SIMPLY AND ADAPT TO THE LANGUAGE OF YOUR STAKEHOLDER.

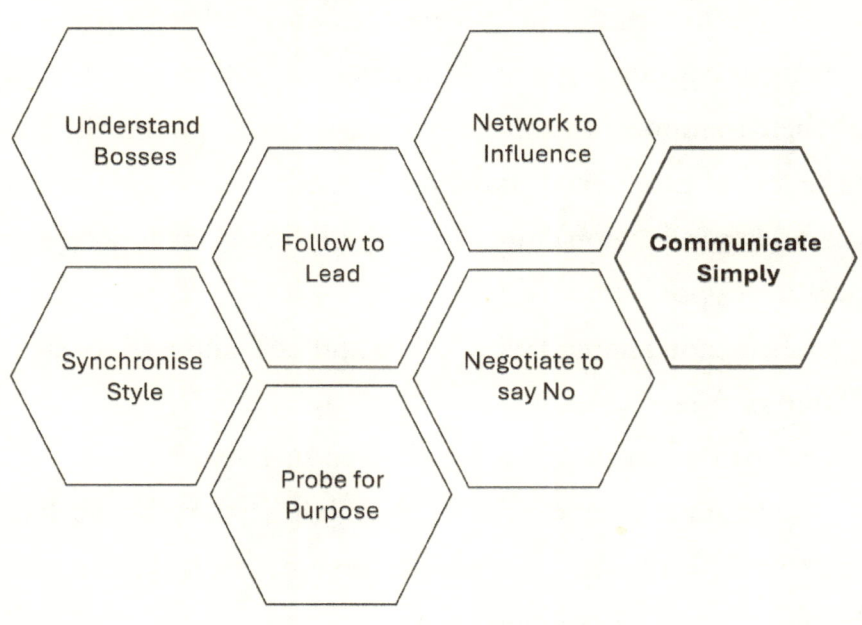

Chris sat with wide eyes and a flushed face. He was talking on his mobile phone and being interrupted continually as he tried to get his point of view across. Eventually, he said goodbye and shook his head slowly.

*Tracy is being impossible. I need help.*

Chris called Olivia, the HR director, and explained the situation.

'Olivia, I tried all that, but she kept going on about the same things. "The money. The flexibility. The company policies. The company values." She is threatening to resign again. We are back exactly where we were before they put her on that merger team. This is urgent. Please, can you come down and see her now?'

'Sorry, Chris. Today is not good for me. As you know, we recently launched the *Balance the See-Saw* employee program, and today a group is going to the range for golf lessons. I must be there to support the initiative. I'm sure you'll handle it. You have lots of experience, and I don't think there is anything I can add that you don't already know. Tell me how it goes. I'm here to support you.'

Chris ended his call with Olivia and said under his breath. 'Thanks. Great support.'

He paused to think. *What am I doing wrong?*

Chris remembered that in their last session, Sonya had reminded him to study the section on escalation. That night, he picked up his book and started reading.

## ESCALATION

There is a limited ability to control things in an organisation. Constrained by finite resources, experience, and authority, managers often encounter scenarios that surpass their immediate sphere of authority.

Central to effective management is the skilful management and communication of risk. When the risk of failure exceeds the benefit of the manager handling the issue, escalation is required. Escalation is not limited to getting help in solving a technical problem. It includes getting support to implement a solution outside the current approved budget or delegated authority. It is also about communicating the broader organisational impact.

Managers who advocate for escalation solely from their vantage point often find themselves disillusioned, failing to garner the anticipated support from stakeholders.

Effective escalation requires clarity. Don't let the complexity of an issue muddy your message. Keep it simple so it's obvious what you need to happen next. A structured approach to escalation mirrors a simple traffic light system.

- The **red** light signifies a crisis necessitating immediate attention and active intervention. The expectation is that the issue is passed onto a higher authority for resolution.
- The **amber** light is a cautionary signal, indicating potential risks or impending danger. This is a warning that things could go wrong, but no intervention is required. The manager doing the escalation still owns the resolution of the issue, and the purpose of the communication is to avoid surprises.
- The **green** light denotes that things are going well. This is a status update to ensure management remains apprised of ongoing activities and achievements.

By adopting this structured approach, organisations can navigate challenges effectively, ensuring timely and appropriate responses while fostering transparency and accountability in decision-making processes.

In the so-called knowledge era, line managers often possess deeper expertise and a stronger grasp of the issue's ramifications than the senior managers they escalate to, upending the conventional hierarchy. The manager must spell out the consequences of the issue in terms that matter to the person they are addressing.

The golden rule is never to escalate a red light issue without preceding it with an amber alert.

'That is so true,' Chris said out loud. He recalled an incident where Fred had not told him how bad the Trueblue account had become. Chris had raised no alarms in his report to senior management, only to discover that Trueblue didn't renew their contract because of ongoing issues that Chris knew nothing about. 'If only I had known sooner,' he blurted out.

Chris continued reading.

Neuroscientists have discovered that our emotions strongly influence our decision-making. Despite the prevalent belief in the workplace that decisions are made solely through dispassionate analysis and logical deduction, empirical findings suggest otherwise. Instead, decisions often stem from emotional impulses, which are subsequently rationalised and substantiated with factual evidence. Decisions are based on emotional feelings and then justified with facts.

The methodology described below uses this understanding of the emotional decision-making journey to escalate issues effectively.

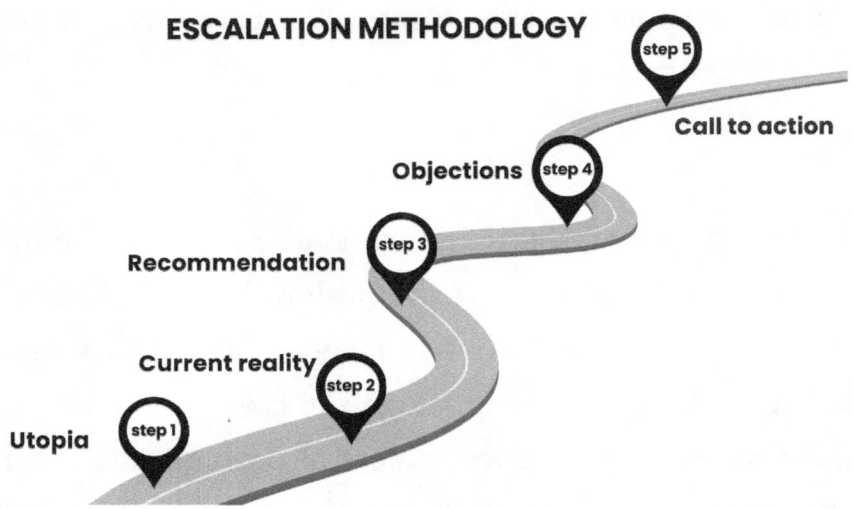

This model uses a framework designed to navigate the intricacies of persuasion, leveraging emotive appeals and logical reasoning to influence decision-makers effectively.

1. Utopia - Begin by illustrating an idealised vision of the desired outcome. Highlight how things would look if the escalation was approved, igniting optimism and enthusiasm within the audience. Vividly portray the resolution of the issue and its attendant benefits to prime individuals to embrace the proposed idea.
2. Current reality - Transitioning to the present reality, describe the repercussions of inaction, laying bare the consequence of maintaining the status quo. The goal is to help the other person

see that the risk of doing nothing outweighs the risk of taking action.
3. Recommendation – Present the recommended solution with the benefits articulated. Quickly getting to the punchline is important, as people are reluctant to *decide* until they know what the conclusion is.
4. Objections – Anticipate and proactively address objections. Decision-makers hesitate to make decisions that could be criticised. By methodically dismantling objections and mitigating perceived risks, the decision-maker feels emotionally safe to support the decision.
5. Call to action – Conclude with a compelling call to action, formalising the decision-making process. Clearly specify what decision is being requested and how it will be communicated.

◆ ◆ ◆

Chris stopped reading and summarised what he had learnt in his mind. So, the colour coded *traffic light system* is a reporting method for escalation, and the *escalation methodology* describes how to escalate. Useful.

Chris reflected on his situation with Olivia, the HR manager, and how he had failed to convince her to get involved with his challenge with Tracy. He had focussed on her availability for the meeting and how it was a problem for *him*, instead of convincing her how critical her involvement was from the company's perspective. *She is the HR director of the company. Of course she cares if we lose an employee. I obviously didn't convince her that her presence was critical to the success of the meeting. How could I have missed that?*

# PRESENTING UPWARDS

Chris and Sonya were catching up casually at the coffee shop before their regular office coaching session. There was a queue of people standing in line for takeaway coffees. The room was a buzz of conversations and laughter blending into one another like white noise.

One of the regulars sat down, motioned to the barista that he wanted his usual order, and tapped his watch to show he was in a hurry.

The barista winked at him, knowing he would get a good tip if he rushed his order through, and the customer grabbed the daily newspaper to read while he waited. He didn't wait long before receiving his coffee.

The cafe owner walked past and said, 'You can't drink

this! Look at the messy latte art. I will get you another one. I am so sorry.'

'Please. I am in a hurry. I don't care about the art. The coffee is fine.'

'Don't be silly. I could not possibly serve you such poor quality coffee. I have a reputation to uphold. I will get you a new one.'

Chris watched the customer groan in frustration as his expedited coffee disappeared behind the counter.

When they got back to the office, Chris opened the conversation.

'I've been asked to present at the monthly operations meeting again. The last one was a disaster. In our session today, can you give me some pointers on how to have more impact?'

'What did you mean by disaster?'

Chris reminded Sonya of what had happened: *not prepared, didn't understand the de*tails, didn't know what I was asking for, ran out of time, no decision.

'Chris, executives want clarity and the big picture. When you walk into a meeting and focus on all the gritty details of your challenges, it's like showing them a duck flapping its feet frantically under the water. Instead of seeing strength, they see something that needs rescuing. Of course, in your world, that paddling is necessary—but that's not what they need to see. When presenting to senior leaders, focus on the direction and speed of the duck—where it's going and how fast. The flapping feet belong in your domain, not theirs. If they see that you're confidently steering the duck toward the same destination they care about, they're far more likely to back you and your ideas. Executives want a high-level summary, not a deep dive into the weeds.'

'Wait a minute. You haven't met our CFO, Quenton. He *loves* details. During the last meeting, he went through my slides with a fine-tooth comb, questioning every data point.'

'Chris, to get Quenton's buy-in at the meeting, you need to go through all those details before the meeting. He is probably the only one in that management team who understands the financial implications. Bosses care that you've looked at the details, but they don't necessarily want to hear them in a meeting. My guess is the other directors look to Quenton for the validity of the business case, and if he gives a thumbs up, they will also give a thumbs up.'

'That is probably true for the others, too,' Chris said. 'I should also get Olivia's buy-in for the people aspects of the decision before the meeting.'

'Chris, technical people tend to overshare. The complexities of technical work are relevant to the designer, as is pedantic accuracy, but to their audience, a less accurate yet understandable version is more helpful.'

'Funny you say that, Sonya—something similar happened to me last night.

'My daughter asked me, "Dad, what colour is a rocket?"

'Naturally, I launched into a full explanation about heat-resistant coatings, reflective properties, and how different aerospace companies use specific colours for branding. I covered everything from thermal shielding to the evolution of paint technology in modern rocketry.'

She blinked, then turned to my wife and asked, "Mum, what colour is a rocket?"

Gwen didn't miss a beat. "Silver," she said.

"Thanks, Mum." And off my daughter ran to grab the silver crayon from her school set.'

'Great example,' said Sonya. 'What is the context of your audience? What do they need to know? Technical people feel uncomfortable simplifying things as it reduces the accuracy of their response, and they are often analytical. To avoid explaining the exceptions and complications, use words like usually, mostly, or frequently. Remember, too, that executives are forward-looking. They look for trends, positive or negative, as they highlight what the outcome will be. So don't show them a table with numbers. Present the data graphically to display the trend you want to illustrate. Show them where the duck is swimming. Also, focus on what they need to know more than what you want to say.

'Make sure you have crystal clear clarity about the purpose of the presentation. Ask yourself, "If everything goes exactly according to plan, what do I hope to achieve?" Write it down in a single sentence.

'When you are presenting, don't make it a lecture. Involve the audience. Ask questions and engage them. Read the room and adapt accordingly. In the book I gave you, *"A Leader's Guide to Managing Up,"* there is a chapter on Constructing the Presentation. Read through that and remember to use the techniques we learnt in escalation. Make it clear to your audience why you are there and what you want them to do.'

Later that night, Chris read the chapter on presentation skills.

## CONSTRUCTING THE PRESENTATION

Clarity begins with simplicity. According to psychological principles, humans prefer information presented in digestible chunks, with sets of three being particularly conducive to comprehension. This phenomenon is evident in various aspects of human cognition, from childhood storytelling featuring trios of characters to the stability of a three-legged stool: three little pigs, three bears and three blind mice. Similarly, the fundamental structure of presentations mirrors this pattern, comprising three essential components. At a minimum: Introduction, Body, and Conclusion. Create additional subsections of three within that structure, where required. Embracing this "rule of three" enhances clarity and coherence, facilitating effective communication and knowledge retention.

- **Introduction** - The art of effective communication starts with gaining attention from the outset. In the introduction, the presenter must harness the power of surprise, fear, or excitement to evoke an emotional response and engender engagement.
- **Body** - Chunking the message into three key points, using the three-legged stool idea, makes it easy for the audience to digest the information. Embracing storytelling techniques and integrating multi-sensory elements cater to varying learning styles, ensuring resonance with the audience's visual, auditory, and kinesthetic preferences. Striking the right balance fosters engagement and enhances message retention, empowering the presenter to deliver information with clarity and resonance.
- **Conclusion** – As the presentation draws to a close, the conclusion serves as the culmination of the narrative, providing closure and reinforcing the key message. Analogous to a sales pitch, the presenter must adeptly "close the sale," compelling the audience to action. A well-crafted

conclusion seamlessly ties back to the impactful opening line, signalling the completion of the presentation journey. Careful preparation ensures alignment with the presentation's objectives, culminating in a persuasive "call to action" that inspires tangible outcomes.

When preparing a presentation, it is essential to consider the three key questions typically on the minds of the audience:

- What is this all about?
- What process is being followed?
- What's in it for me?

Craft the content in reverse order of presentation delivery. Commencing with the conclusion allows the presenter to crystallise the primary purpose of the presentation, guiding the subsequent development of the body and introduction. This approach fosters coherence and alignment, minimising the risk of overwhelming the audience with excessive key messages.

Structure the body around the key points of the conclusion, to avoid side issues confusing the audience.

Finally, preparing the introduction last lets presenters maximise impact, leveraging insights gleaned from the conclusion and body to craft an attention-grabbing opening statement.

Effective presentation delivery encompasses not only the content conveyed but also the demeanour and presence of the presenter. It is imperative to project a polished and professional image, beginning with appropriate attire tailored to the occasion. In the digital realm, attention to one's online presence and background is equally vital, as it contributes to the overall impact and perception of credibility and professionalism.

When speaking, deliberate pacing of speech is key to maintaining audience engagement and coherence. Speaking slowly mitigates the tendency to fill pauses with distracting verbal tics such as "ums" and "ers." Speak with confidence and clarity to cultivate an atmosphere of authority and expertise. This will enhance audience receptivity and comprehension.

Do not start with an apology. Audiences are inherently supportive and eager for the presenter to succeed; thus, setting low expectations undermines their confidence and detracts from the overall experience. Instead, embrace a mindset of passion and enthusiasm, infusing the presentation with energy and vitality. Executives and audiences alike value informative and engaging presentations that captivate attention and leave a lasting impression.

Chris closed his book, and a broad smile came across his face as he thought about the upcoming senior management meeting presentation that was scheduled for the end of the month. *You have no idea what you are in for.*

# EXECUTIVE BRIEFING SUCCESS

Chris walked into the meeting room. Rob sat at the head of the table. Gabriella, Olivia, Quenton, and the other managers spread themselves around the perimeter. A half-eaten plate of pastries sat tantalisingly in the middle of the table with empty coffee cups strewn about untidily. The room smelled a little musty. *Deja Vu.*

'Sit, Chris. And help yourself to a pastry. They are delicious.'

'So what's next on the agenda?' Rob asked.

Quenton spoke up. 'Chris is going to update us on the progress of the latest product development plans. We couldn't come to any decisions last month, so we asked Chris to present again because we need to decide on approving increased funding for the software.'

Chris jumped to his feet.

'Imagine if I told you our new product was ready for release *and* was under budget. Our current reality is that we are far from releasing it, and our project costs will exceed estimates by 30%. The good news is, I have a solution.'

Chris walked over to the laptop and presented a glitzy slide of a new software module called *Modex*.

'*Modex* is the answer. This module is off-the-shelf and provides 90% of what we need. Our software developers can easily add any missing functionality because it uses publicly available *open-source* software.'

The other managers turned to Gabriella. 'Is this true?'

'Yes, absolutely. Chris took me through the whole concept, and it's a winner.'

Chris took back the floor. 'Now you may be wondering, how much will this cost? If you look at the handout I have circulated, I completed a business plan that shows the payback period on the initial investment is less than 6 months, and it will bring us in under budget overall.'

Rob turned to Quenton, the CFO. 'Have you seen this?'

'Yeah, the numbers stack up. I put Chris's data through our standard evaluation tool, and his estimates are very conservative.'

One by one, Chris addressed all the potential pitfalls of the new module.

Finally, Chris concluded. 'Every day we delay our decision,

it costs us money. Have I got your go-ahead to start the process and get quotations to purchase this software?'

Rob raised his eyebrows and slowly looked around the room as he received unanimous approval.

'Chris. We approve. Good job. Hey, do us a favour and take these pastries to the company kitchen. I'm afraid we'll be rolling out of this meeting if they stay here any longer.'

# GOTCHA

Chuck was sitting talking to Tracy at the coffee shop.

'Hold your ground with these people, Tracy. They are exploiting your youth. Don't let Chris get away with this.'

'It's awful, Chuck. It's like he doesn't trust me and wants to check up on me 24/7. The other day, I was late back from lunch and missed a customer meeting. Doesn't everyone forget sometimes? Chris tried calling my mobile, but I always turn it off during lunch. He told me he doesn't mind me going off site to lunch, but to tell him where I am during work hours so he knows how to get hold of me. What a micromanager.'

'That's unbelievable, Tracy. Does he think he is a sergeant major in the army? There are labour laws that protect you from bullies like him. I have a friend who is an HR attorney.

I'll get them to help you draft an email requesting greater flexibility. You watch. Chris's reply will help us nail him.'

'Thanks, Chuck. I've already emailed senior management complaining about him. I have threatened to resign and expected they would move heaven and earth to keep me, but there has been no response so far. I might need to wave an actual resignation letter in their faces, then we'll see action. I appreciate your support. You've been a rock to me.'

Chuck's mobile went off. 'Sorry, Tracy. I've got to run. Important company business. See you around.'

Tracy watched Chuck walk off. He had a worried expression on his face. She immediately finished her coffee and headed back to the office. She noticed Chuck walking past the office. She paused to watch him. Further down the road, a mystery black BMW with dark tinted windows stopped, and Chuck got in. *That's odd.*

Back in the office, Rob was on the phone with Olivia. 'I have received a complaint about Chris from one of his team members, Tracy. She is threatening to resign unless we change her reporting structure.'

'Tracy. Yes. I am afraid she has become a detractor in the business. She is very bright, and we thought she had a lot of promise when we hired her.'

'I know. I put her on the merger team last year,' said Rob.

'Well, I think it went to her head. She keeps complaining about Chris but says nothing concrete. She keeps saying, "I don't want to get him into trouble," and will not be specific about her complaints. Chris brought the issue to my attention a while ago, but initially, I didn't realise how serious it was. Since then, he has explained to me the business impact of her behaviour. I have spent time trying to coach her, but she is very self-assured and not open to learning. She excels in things she enjoys doing, but her general reputation around the business is poor. Chris says he has even had complaints from customers about her. I recommend that if she goes ahead with a resignation letter, we accept it.'

'I agree,' said Rob. 'We need team players in our organisation, and Tracy does a lot of complaining but does not seem to have any suggestions or solutions. I'll leave it with you to respond.'

Rob ended the call and dived into reviewing a customer proposal, and was deep into the financial analysis of the deal's profitability when his phone rang.

'I see— of course — I understand. Thanks for telling me. Yes, you have a good day too.'

Rob immediately called Olivia.

'We need to meet. Urgently.'

'About Tracy?'

'No, this is far more serious.'

Olivia went straight up to meet Rob in his office. She listened as Rob explained the situation.

'I received a call from one of our competitors. We are in discussions with them about another merger. In the interests of full transparency, they told me that a rogue sales manager in their organisation had been caught redhanded, paying one of our sales team a commission to deliberately lose a deal. They found an entire file of confidential information from our company that had been leaked to them by our sales lead.

'Olivia, the sales lead is Chuck.'

'Chuck? I can't believe it. He's such a nice guy.'

'Well, Chris has been warning me about him, but I thought it was a personality clash. Chuck tricked us all. Unfortunately, many sociopaths and psychopaths can hide in organisations. The personalities of narcissists and sociopaths make it easy for them to be manipulative and abuse the organisational power they are given.'

'So true, Rob. What a day. I still can't quite believe it.'

# TRANSFORMATION COMPLETE

An announcement was sent out company-wide telling everyone that Chuck had left the business with immediate effect *to pursue personal interests.*

A few months passed. Thomas recovered to full health and returned to his old role as Director of Engineering, but kept a low profile. Tracy resigned, telling anyone who would listen about her dramatic action. She was shocked when her resignation was politely accepted.

Chris had continued to meet with Sonya, though now it was once a month. This time, they had agreed to catch up in person at the coffee shop.

Sonya spotted Chris waiting out front and gave him a wave.

'Fancy a walk? It is such a lovely day. We can talk as we walk.'

'So long as I can take my coffee with me,' Chris said.

Sonya ordered two takeaway coffees, and they headed towards the park.

'Chris, have you thought about what legacy you want to leave behind in your role?'

'Legacy? I am just starting my senior management career. Why would I think about my legacy?'

'Chris, every time you interact with someone, you leave behind something of yourself. We are all connected. The way you deal with people leaves a lasting impression on them. Think about what you want to be known for in 5 or 10 years. Living up to that standard will guide you in all your decision making. You have grown so much in the past few months. The secret is to keep growing. In golf, your handicap keeps you competing against your own standard. As you improve, your handicap drops so you never get complacent. That's the standard to aspire to - your own evolving standard.'

'I like that analogy, Sonya. I think previously I was aiming to achieve success in my area without any consideration for those around me. My lesson in managing up is, it starts with supporting the organisation. I used to think that managing up was about kissing up. I've realised that it is not about me trying to look good with my bosses for my personal gain but to increase my influence. It is also not about me being subservient with my bosses. It is about finding that win-win outcome.

'I've realised I can't change my bosses, but I can change the way I interact with them. Now that I'm adapting my style to suit the person I'm dealing with, I'm finding I have greater influence. I've also realised the importance of stepping into my boss's shoes before making judgments, and becoming more aware of my biases.

'I think my biggest lesson is to engage in their world, not mine. Thinking of who they have to deal with - the two up-two-across rule has really helped me understand what is *actually* required.

'Last, I've realised I need to be more positive. Not toxic positivity. That would annoy the staff, but I need to avoid painting a dark, gloomy picture because of my own risk aversion. The future is made-up and imaginary, so it might as well be optimistic. An optimistic outlook often leads to better outcomes and that's exactly what bosses really, really want.

'The legacy I want to leave is to make a difference in all my interactions, not only with the team I lead.'

'That's great, Chris. You have certainly come a long way in your journey. The beauty of progress in managing up is that progress comes mainly through a different mindset. All the actions that follow happen based on thinking differently about your role. It's as though there's a switch setting in your head that needs to be flicked, and then everything looks different. In our coaching session next month, we may talk about looking at a new leadership development area to tackle.

Like golf, we need to look at which areas of your game need improvement and focus on those to improve your handicap.'

Chris chuckled and nodded. 'Makes sense,' he said. He thanked Sonya and headed back to the office. As he stepped inside, his phone buzzed with a text from Rob: 'Any chance you can pop in and see me?'

'I'm free now if you are.' Chris replied.

Rob was free, so Chris headed off to his office.

'Hi Chris, please take a seat and close the door behind you. The dust has settled from the merger and the product development work is going well. Our biggest challenge is expansion into these new markets. Our existing sales team is used to selling to technical buyers, and as you know, we have been unsuccessful in replacing Chuck. The new products require a different distribution channel and sales approach. As you were saying the other day, we have to develop a more strategic approach to technical sales. We have found a company that has an excellent distribution channel, but their products are outdated. We are looking to create a joint venture company.

'Chris, Thomas spoke to me and has decided to go back to his roots as a Project Manager. Gabriella has recommended you for the Engineering Director's role. If we offered it to you, would you accept?'

'Absolutely.' said Chris.

'Great. I will work with Olivia to get a formal offer out to you with the financial implications of your promotion. And Chris—'

'Rob?'

'We would like you to be on the due diligence team to assess the feasibility of this new merger. Remember, this is top secret, so you can't talk to anyone about it. Are you in?'

'Yes, Rob. I'm in.'

# A NOTE FROM THE AUTHOR

While the characters in this book are fictional, many of the workplace stories told through their dialogue are based on real events. During the beta review process, some readers felt a few stories or conversations seemed far-fetched—but in truth, I've softened the original versions to protect the real people involved. As they say, sometimes fact really is stranger than fiction.

    I wrote this book to complement my leadership training on *Technical Team Leadership* and *Managing Up and Across*, and to offer practical insights to anyone involved in leading technical people. To support this, I've included a summary of key themes and where to find them throughout the book for easy reference.

# TOPIC INDEX

Key Themes - takeaways                                          Chapter

1. Introduction to managing up                                        3
2. Understanding the new role of bosses in a post-Knowledge era (Lesson 1)        5
3. Influencing New bosses                                             6
4. Alignment of style for influence (Lesson 2)                        7
5. Importance of following to lead (Lesson 3)                        10
6. Authority versus influence (Lesson 4)                             11
7. The 2-up-2-across rule to understand what is REALLY required (Lesson 4)       11
8. Multiple bosses - dealing with work overload and conflicting priorities (Lesson 4)   11
9. Establishing strategic allies through networking to overcome lack of authority to get things done (Lesson 5)    13
10. Saying no using negotiation levers and the 'yes, and' strategy (Lesson 6)    14

11. Escalation reporting - traffic light system (Lesson 7) — 15
12. Escalation methodology (Lesson 7) — 15
13. Presenting upwards - constructing the presentation (Lesson 7) — 16
14. Bad and abusive bosses — 18
15. Managing up lessons — 19

# APPENDICES

# APPENDIX 1 - INFLUENCING STAKEHOLDER SELF-SURVEY

For each aspect below, assess whether you need improvement.

1. I successfully manage up, across and out from the business.
2. I have influential relationships with key bosses, other departments, customers, suppliers and business owners.
3. I match and adapt my style to that of my key stakeholders.
4. I show by my attitude and behaviour that I am a good follower of senior leadership.
5. I know which bosses are key to my overall effectiveness and long-term value to the business.
6. I have identified strategic allies inside and outside the organisation that I need to network with to achieve my business goals.

7. I have an influence plan for my key stakeholders that matches their goals and aspirations and includes regular review and relationship building.
8. I know how to say no, and deal with organisational conflict.
9. I know how to escalate issues to get an urgent and positive response.
10. I know how to do business presentations effectively to all levels in the business.

# APPENDIX 2 - KEY INFLUENCER PLAN - SELF-REFLECTION

*"You can't influence someone you don't understand."*

This worksheet is designed to help you strategically strengthen your relationship with your boss or key stakeholders by stepping into their world and aligning mutual goals. Use it as a practical guide to apply the core principles from this book.

**1. Understand Their World**
    a. What are your boss's goals?
- Stated work priorities:
- *Unstated* priorities (e.g., reputation, job security, career progression):
- Personal drivers or motivations (if known):

    b. Where might your goals and theirs be in conflict?
- Conflicts:
- Your strategy to align or reframe the differences:

    c. If you don't know their goals—what's your plan to find out?
- Specific questions to ask or behaviours to observe:
- Stakeholders or colleagues who may help illuminate this:

**2. Clarify what You Need**
   a. What do you expect from your boss?
      - Support needed (e.g., coaching, visibility, feedback):
      - Resources or decisions only they can provide:
   b. Have you clearly communicated these expectations?
      - ☐ Yes  ☐ No   If no, what actions will you take to initiate this conversation?
   c. Are your expectations realistic based on their role, capacity, and influence?
      - ☐ Yes  ☐ No   Adjustments needed:

**3. Clarify what They Need**
   a. What deliverables or results does your boss need from you?
      - Core responsibilities and metrics:
   b. What do *they* owe to their own leaders or stakeholders?
      - Known pressures from above:
      - How can you make their life easier and help them look good?
   c. What top challenges or pain points does your boss face and how could you proactively help with these challenges?

### 4. Strengthen Communication

a. Do you have a regular forum for communication?
- ☐ Yes ☐ No  If not, schedule or propose one. Frequency: _____
   Preferred format: ☐ Email ☐ Video/Phone ☐ Face-to-face

b. Have you agreed on how and when to escalate issues?
- ☐ Yes ☐ No  Describe what needs to be agreed upon:

c. Is your crisis their crisis?
- ☐ Yes ☐ No  How can you create shared visibility of problems?

d. Actions to improve two-way communication:

### 5. Help Them Win

a. What actions can you take to help your boss succeed?

b. How do those actions also help you achieve your goals?

### 6. Investing in the Relationship

a. What behaviours, habits, or assumptions will you change?

b. What do you plan to change to improve your relationship with your boss?

**7. Action Plan**
   a. What will you do differently having read this book?
      - Start doing ….
      - Stop doing ….
      - Keep doing ….

# RECOMMENDED READING AND REFERENCES

1. Barrett, Lisa. *How emotions are made.* Pan MacMillan, 2017
2. Berne, Eric. *Games people play.* Harper Collins, 1964
3. Gallo, Amy. *Getting along: How to get on with anyone (even difficult people).* Harvard Business Review Press, 2022
4. Kahneman, Daniel. *Thinking, Fast and Slow.* Farrar, Straus and Giroux, 2011.
5. Peters, Steve. *The Chimp Paradox: The Mind Management Programme to Help You Achieve Success, Confidence and Happiness.* Vermilion, 2011.
6. Scott, Kim. *Radical Candor.* St Martin's Press. 2017
7. Stone, Douglas. & Heen, Sheila. *Thanks for the Feedback: The Science and Art of Receiving Feedback Well.* Viking Press, 2014
8. Stone, Douglas, Patton, Bruce & Heen, Sheila. *Difficult conversations: How to discuss what matters most.* Penguin Books, 2010

9. Patterson, Kerry. *Crucial Conversations*. McGraw-Hill, 2011
10. Voss, Chris, & Raz, Tahl. *Never Split the Difference: Negotiating As If Your Life Depended On It*. Harper Business, 2016

# ABOUT THE AUTHOR

Trevor Manning is a consultant, author, and trainer known for helping technical professionals communicate their expertise so it drives real business impact. With a career that bridges deep technical mastery and executive leadership, Trevor brings a rare combination of insight to his work. As a microwave radio planning expert who advanced to C-level roles, he understands firsthand how technical experts can find themselves sidelined—not because of a lack of value, but because they struggle to translate their ideas into the language of business. Trevor learned how to bridge that gap, and now dedicates his work to helping others do the same.

His career includes roles such as Chief Engineer at a major electric utility, Technical Director at an American manufacturing firm, and COO in both a software startup and a private telecommunications company. Trevor has worked across South Africa, the UK, Italy, the US, and Australia—gaining broad international experience in business culture and practices.

This is Trevor's tenth book. His *Microwave Radio Transmission Design Guide* (published by Artech House) is widely regarded as a classic in the field of microwave engineering and remains a trusted reference for industry professionals.

Trevor has trained thousands of business professionals through his leadership courses at Oxford University and University of Wisconsin, Madison.

Trevor lives in Brisbane, Australia.

# OTHER LEADERSHIP BOOKS BY TREVOR MANNING

*Help! What's the secret to leading engineers?* (2017)
*'Help! I need to master critical conversations.* (2018)
*'Help! They made me the Manager.* (2021)

These books are available from Amazon.

www.ingramcontent.com/pod-product-compliance
Lightning Source LLC
Chambersburg PA
CBHW022016290426
44109CB00015B/1187